The role of households in solid waste management in East Africa capital cities

The role of households in solid waste management in East Africa capital cities

Aisa Oberlin Solomon

Environmental Policy Series – Volume 4

Wageningen Academic
Publishers

This work is subject to copyright. All rights are reserved, whether the whole or part of the material is concerned. Nothing from this publication may be translated, reproduced, stored in a computerised system or published in any form or in any manner, including electronic, mechanical, reprographic or photographic, without prior written permission from the publisher: Wageningen Academic Publishers P.O. Box 220 6700 AE Wageningen The Netherlands www.WageningenAcademic.com copyright@WageningenAcademic.com

ISBN: 978-90-8686-191-0
e-ISBN: 978-90-8686-747-9
DOI: 10.3920/978-90-8686-747-9

The content of this publication and any liabilities arising from it remain the responsibility of the author.

First published, 2011

© Wageningen Academic Publishers
The Netherlands, 2011

The publisher is not responsible for possible damages, which could be a result of content derived from this publication.

Preface

Solid Management is a concern in East African capital cities. The absence of managing solid waste is a serious problem. An ever bigger concern is the growing quantities of waste that are generated at households level in informal settlements. In most cases proper safeguard measures are largely ineffective or not in place at all. Moreover, unsafe disposal of waste in the region is coupled with poor hygiene. There is no doubt that East African capital cities need to formulate effective ways to manage their waste.

This book is a result of PhD research within the framework of the PROVIDE project funded by INREF and carried out in East African capital cities (Dar es Salaam, Nairobi and Kampala). The work was carried out at the Environmental Policy (ENP) group with professor Gert Spaargaren and Dr. Peter Oosterveer as promotor and co-promotor. The focus of the research was to highlight the role of households in the production and management of domestic solid wastes.

Inevitably numerous persons and organizations have contributed to this research since it began in November 2006. Each contribution has in some way enriched the process and facilitated my ability to bring the research to a successful conclusion. I gratefully acknowledge the support of all those who have assisted me in this endeavour but I will also make specific mention of those persons whose contributions have been especially significant to my work.

More importantly and admittedly, I deeply appreciate the financial support offered to me by the INREF and the Environmental Policy Group under the PROVIDE project during the period of my study. I am also grateful to the study leave granted to me by my employer Dar es salaam Institute of Technology (DIT).

I am heartily thankful to my promotor prof. Gert Spaargaren and co-promotor dr. Peter Oosterveer whose encouragement, guidance, comments and support from the initial to the final stage enabled me to develop an understanding of the subject. Without their dedication, this work would not have become a reality. Their advice throughout the work, as well as their constructive criticism have influenced my thinking. Through you I learnt to believe in my work and myself. I would like to give special appreciation to prof. Gert Spaargaren for who accepted me as his Ph.D. student without any hesitation. He offered me so much advice, patiently supervising me, and always guiding me in the right direction. As a result, research life became smooth and rewarding for me.

I want to express my deeply felt thanks to PROVIDE coordinator Dr. Gabor Szanto, for his constant guidance, encouragement and for listening to my ideas. I also appreciate the advice and support from PROVIDE project partners from East Africa; Dr. Shaaban Mgana from ARU, Dr. James Okot-Okumu from Makerere University Institute of Environment and Natural Resource, Kampala (Uganda); and Dr. Caleb Mireri from Kenyatta University, Department of Environmental Planning and Management, Nairobi, Kenya. We worked very closely together throughout. A special thanks goes to Dr. Shaaban Mgana for his instructions and assistance during my field work.

I enjoyed the company and the logistical support given by the other members of Environmental Policy Group, their cooperation and enthusiasm were essential. I thank you all for the strong cohesion. Corry Rothuizen deserves a special word of thank, for making sure that I had pleasant working environment and for taking care of the necessary logistics.

My fellow scholars in the PROVIDE project are Judith Tukahirwa, Christine Majale, Mesharch Katusiime, Sammy Letema, Thobias Bigambo, Fredrick Salukele, Richard Oyoo. I am grateful for the friendship, laughter, care, support and the valuable time we had together. The journey became easier as we travelled together. I am grateful to Judith Tukahirwa for being a constant source of inspiration and great help.

My time at Wageningen was made enjoyable due to the many friends and groups that became a part of my life. I am grateful to my roommates Christine Majale, Dorien Korbee and Hilde Toonen. I am also thankful to Harry Dabban Barnes, and Leah Ombis for being a great company.

I will always treasure the social interactions with the past and present student community at Wageningen University. I have fond memories of the get-together parties with Tanzanians and the wider East African Community which offered me desired moments of refreshments and relaxation.

I am deeply indebted to the students' chaplaincy in Wageningen, ICF International and Amazing Grace Parish for their immeasurable spiritual sustenance. Worshiping at ICF was a wonderful experience. I cherished the prayers, Bible studies and support, and the friendship with my Christian brothers and sisters. I treasured all precious moments we shared, I would like to thank you so much.

My deepest gratitude goes to my family for their persistent love and support throughout my life; this dissertation would simply be impossible without them. I am indebted to my father, Rev. Oberlin Solomon, for his care and love. He worked hard to support the family spiritually and spared no effort to provide the best possible environment for me to grow up and attend school. Although he is no longer with us, he is forever remembered. I am sure he shares our joy and happiness in heaven. To my loving mother, I have no suitable word that can fully describe her everlasting love to me. Her constant prayers and unconditional love have been my greatest strength. Mama nakupenda (Mother, I love you).

Last but not least, I am greatly indebted to my devoted husband Dr. Edward E. Sawe. For intimate advice and personal support while doing this study, I relied almost entirely on my dear husband. He has been so inspiring for the entire period of my study. To my children Nsia, Elisha and Aluseta, I owe you a lot. They form the backbone and origin of my happiness. Their love and support without any complaint or regret has enabled me to complete this PhD programme.

Table of contents

Preface   7

Abbreviations   15

Chapter 1.
Introduction   17
  1.1 The problem of household waste   17
  1.2 Research background   18
  1.3 Problem specification   18
  1.4 Relevancy of the study   19
  1.5 Outline of the thesis   20

Chapter 2.
An overview of solid waste management in Dar es Salaam city   21
  2.1 Introduction   21
  2.2 Description of Dar es Salaam city   22
  2.3 Description of the solid waste management system   23
    2.3.1 The situation from 1982 to 1992   23
    2.3.2 Initiatives to improve solid waste management in Dar es Salaam city   24
  2.4 National policies and the regulatory and institutional frameworks relevant for
      solid waste management   30
    2.4.1 Solid waste management policy framework in Dar es Salaam   30
    2.4.2 The Tanzanian legislative and regulatory framework for solid waste management 32
    2.4.3 The institutional arrangements for SWM in Dar es Salaam   35
  2.5 Existing solid waste management practices: some facts and figures   36
    2.5.1 Total waste generation in Dar es Salaam city   37
    2.5.2 Solid waste storage   38
    2.5.3 Collection of solid wastes   38
    2.5.4 Disposal of waste   41
    2.5.5 Resource recovery   42
  2.6 The waste stream and waste disposal practices   43
  2.7 Resources available to the Dar es Salaam City Council for solid waste management 45
    2.7.1 Financial resources   45
    2.7.2 Technical resources   46
  2.8 Conclusions   47

Chapter 3.
Understanding and improving household-waste management in East
African capital cities: conceptual framework   49
3.1 Introduction   49
3.2 Integrated solid waste management   49
3.3 The concept of Modernized Mixtures Approach in solid waste management   51
3.4 The concept of households   54
   3.4.1 Households as (key) elements and stakeholders in solid waste management
   chains   56
3.5 Conceptualizing the roles of households in domestic solid waste management
   chains   57
   3.5.1 Households as key units of analysis   59
   3.5.2 Solid waste management infrastructures   60
   3.5.3 Modernized Mixtures Approach criteria: ecological sustainability,
      accessibility and flexibility   61
3.6 Research questions   62

Chapter 4.
Research methodology   63
4.1 Data collection methods   63
4.2 Methods of primary data collection   65
   4.2.1 Reconnaissance visit   65
   4.2.2 Interviews with key informants   65
   4.2.3 Focus group discussion   67
   4.2.4 Direct observation   69
   4.2.5 Secondary materials   69
   4.2.6 Quantitative data collection   70
4.3 Waste characterization study   74
4.4. Training of research assistants   75
4.5 Research reliability and validity   76

Chapter 5.
Household solid waste characteristics   77
5.1 Introduction   77
5.2 Conceptual framework   78
5.3 Methodology   81
5.4 Waste characteristics   82
   5.4.1 Per capita waste generation   82
   5.4.2 Physical composition of waste   83
5.5 Factors that affect solid waste characteristics   87
   5.5.1 Demographic/socio-economic factors   87
   5.5.2 Lifestyle related activities   92
5.6 Conclusion and discussion   94

Chapter 6.
Household waste handling   **97**

6.1 Introduction   97
6.2 Conceptual framework   98
6.3 Methodology   100
6.4 Current waste management practices as applied by households   100
    6.4.1 Waste storage   101
    6.4.2 Waste separation at the household level   102
    6.4.3 Composting at the household level   104
6.5 Waste flows travelling from the household to the transfer station   104
    6.5.1 The formal collection systems   104
    6.6 Alternative disposal methods used by households   108
    6.6.1 Informal systems for waste handling   108
6.7 Roles of different household members in the practice of waste management   111
6.8 Conclusion and discussion   112

Chapter 7.
Households as service recipients in solid waste management chain   **115**

7.1 Introduction   115
7.2 Data collection methods   118
7.3 Formal and informal stakeholders and their roles in domestic solid waste
    management   119
    7.3.1 The role of municipal authorities in domestic solid waste management   119
    7.3.2 The role of waste contactors in domestic solid waste management   120
    7.3.3 The role of informal waste pickers in domestic solid waste management   122
7.4 The relationship between households and other key stakeholders in domestic
    solid waste management   123
    7.4.1 The relationship of households with formal stakeholders   124
    7.4.2 Informal relationships of householders with other stakeholders   125
    Household and their relations with waste pickers   126
7.5 Municipal assistance to stakeholders in domestic solid waste management   128
7.6 Householders' perceptions and assessment of the relationships with formal and
    informal waste management actors   130
    7.6.1 Perceptions and assessments as obtained from the questionnaire survey   130
    7.6.2 Perceptions and assessments as obtained from the focus group discussion   133
7.7 Conclusion   137

Chapter 8.
Households and domestic waste management in comparative
perspective: some findings from Kenya/Nairobi and Uganda/Kampala   **139**
8.1 Introduction   139
8.2 Basic characteristics of the three cities   139
8.3 Research methodology   140
8.4 Solid waste management in Nairobi and Kampala   142
    8.4.1 The solid waste management organization in Kenya/Nairobi   142
    8.4.2 The solid waste management organization in Kampala/Uganda   142
8.5 Empirical findings in Nairobi   143
    8.5.1 Practices of waste handling in Nairobi: collection and transfer 143
    8.5.2 Households perceptions in Nairobi   145
8.6 Empirical findings in Kampala   146
    8.6.1 Households as waste handlers in Kampala: collection and transfer 146
    8.6.2 Household perceptions in Kampala   147
8.7 Comparing Dar es Salaam, Nairobi and Kampala on a selected number of solid
waste management aspects   148
    8.7.1 Solid waste management organization   149
    8.7.2 Service provision to households   149
    8.7.3 Households perspectives   151
8.8 Lessons learned from the comparative analysis   152
8.9 Conclusion   154

Chapter 9.
Conclusion and discussion   **157**
9.1 Introduction   157
9.2 The Modernized Mixture Approach as a framework for assessing domestic solid
    waste practices in informal settlements in Dar es Salaam   158
9.3 Households as waste generators   159
    9.3.1 The per capita daily waste generation   160
    9.3.2 The composition of domestic wastes   161
    9.3.3 Relationship between demographic variables and waste characteristics (per
        capita waste generation and waste composition)   162
9.4 Households as waste handlers   163
    9.4.1 Solid waste management practices applied by households   163
    9.4.2 Management of waste transfer stations   167
    9.4.3 Alternative disposal methods   168
9.5 Households as service recipients in solid waste management chain   168
    9.5.1 The relationships of households with other stakeholders in domestic solid
        waste management   169
    9.5.2 The role of municipal authorities in assisting stakeholders in the waste
        chain: the role of public-private parnerships   170
    9.5.3 Households' perceptions and assessments of waste service provisioning 171

9.6 Reflection on the conceptual framework: flexibility, accessibility and ecological
   sustainability   173
9.7 Main conclusion   174
9.8 Some suggestions for future research   176

References   **179**

Appendices   **187**

Summary   **205**

About the author   **209**

Abbreviations

ADB	African Development Bank
ARU	Ardhi University
CBO	Community based organization
CCM	Chama Cha Mapinduzi
CLN	Electrical and General Contractors Ltd.
CSRC	Crisis State Research Centre
CWG	Collaborative Working Group
DCC	Dar es Salaam City Council
DeSaR	Decentralized Sanitation and Re-use
DIT	Dar es salaam Institute of Technology
DSM	Dar es Salaam
EMA	Environmental Management Agency
EPA	Environmental Protection Agency
ERC	Environmental Research Consultancy
GCLA	Government Chemistry Laboratory Agency
HNS	Hananasifu
HSW	Household Solid Waste
IDRC	The International Development Research Centre
ILO	International Labour Organization
ITC	International Institute for Geo-Information Science and Earth Observation
INREF	Interdisciplinary Research and Education Fund of Wageningen University
ISWM	Integrated Solid Waste Management
JICA	Japan International Cooperation Agency
KCC	Kampala City Council
KENAO	Kenya National Audit Office
KIWODET	Kisutu Women Development Trust
KIMODA	Kinondoni Moscow Development Association
KLH	Kilimahewa
KMC	Kinondoni Municipal Council
MDG	Millennium Development Goals
MKR	Makangira
MLHSD	Ministry of Lands and Human Settlements Development
MMA	Modernized Mixtures Approach
NBS	National Bureau of Statistics
NCC	Nairobi City Council
NEMA	National Environment Management Authority
NEMC	National Environmental Management of Tanzania
NGO	Non-governmental organization
NIGP	National Income Generation Programme
NL	The Netherlands

PMO	Prime Minister's Office
PROVIDE	Partnership for Research on Viable Environmental Infrastructure Development in East Africa
RCC	Refuse Collection Charges
SKAT	Swiss Centre for Development Cooperation in Technology and Management
SACCO	Savings and Credit Cooperative
SCP	Sustainable Cities Programme
SDP	Sustainable Dar es Salaam Project
SWM	Solid waste management
SUDP	Strategic Urban Development Plan
TS	Transfer Station
TTM	Tua Taka Makurumla
TZS	Tanzanian Shilling
UMP	Urban Management Programme
UN	United Nations
UNCED	United Nations Conference on Environment and Development
UNCHS	United Nations Centre for Human Settlements
UNDP	United Nations Development Programme
UNEP	United Nations Environment Programme
UN-HABITAT	The United Nations Human Settlements Programme
USD	United States Dollar
USEPA	U.S. Environmental Protection Agency
UWEP	Urban Waste Expertise Programme
WASTE	Advisers on Urban Environment and Development
WELL	Water and Environmental Health at London and Loughborough
WB	World Bank

Chapter 1.
Introduction

1.1 The problem of household waste

The ever-increasing quantities of household wastes are a growing environmental problem in urban centres in both developed and developing countries around the world (Barr *et al.*, 2001). As a result, domestic solid waste poses a complex challenge for environmental policy (Barr, 2007). The blueprint for worldwide sustainable development agreed by national leaders in Rio de Janeiro in 1992, Agenda 21 (UNCED, 1992), already highlighted waste from domestic sources as a major barrier to achieving environmental sustainability in the 21st century. The problem of managing solid waste in the urban areas must be seen in the wider context of problems caused by rapid urbanization. In Africa a growing number of cities face the challenge to provide their populations with adequate water supply, sanitation and solid waste services, because of the rapid rate of the urbanization process. United Nations projections estimate that the urbanization rate will increase from 24% in 2005 to 38% by 2030, with more than 20 million Tanzanians living in urban areas (United Nations, 2006).

The fast increasing amounts of waste have to be processed, while in Africa, the management capacities available for this purpose are often poor and inadequate, especially in low-income areas. Most municipal authorities in developing countries have failed to provide their expanding populations with adequate services for managing solid waste as well as for providing water and sanitation (Abduli, 2007). Also in East Africa and in its capital cities in particular, urban solid waste management poses a serious environmental problem. The fast growing quantities of domestic wastes constitute an enormous challenge for the local authorities. In order to improve their strategies for managing domestic solid wastes, a better understanding of both the technological and managerial aspects or dimensions is needed. While various reports, projects and policy documents on the subject of solid waste management are available, the role of households as primary producers of solid wastes tends to be overlooked. Domestic actors tend to be neglected, both in their role as waste handlers as well as in their role as stakeholders and potential contributors to solving the problem. This oversight seems to be the consequence of inadequate assessment and conceptualization of the role of households in urban solid waste management.

To make up for these shortcomings, this study illustrates how households are conventionally conceptualized in the literature and what the main challenges are when trying to understand their potential contribution for improving the situation in East Africa. It develops a new conceptual framework for analyzing the role of households in solid waste management in East Africa's capital cities. This conceptual framework is derived from the Modernized Mixture theory. In this new approach, households and their practices of waste management are at the centre of attention. Domestic routines for handling wastes are described and analyzed both for their technical and their social dimensions. By focusing on households in informal settlements in particular, the study contributes to the theory of the Modernized Mixture Approach in a specific way.

By elaborating upon both the technical and social aspects of domestic solid wastes and the role of householders in producing and handling these wastes, this present study adds to the scanty body of scientific knowledge sustainable waste management by householders in East African countries. The knowledge generated with respect to technical and social dimensions can be used in the future by researchers and policy makers in their search for more effective and sustainable SWM policies both in East Africa and comparable situations elsewhere in the world.

This chapter is organized as follows: in section 1.2, a brief background of this research is given. Section 1.3 introduces the problem of household waste management, and highlights in particular domestic solid waste management problems in informal settlements in East African capital cities. Section 1.4 presents the societal relevancy of the study, indicating the overall objective which is linked to the realization of the Millennium Development Goals; section 1.5 presents the outline of the thesis.

1.2 Research background

This study is undertaken within the framework of the project Partnership for Research on Viable Environmental Infrastructure Development in East Africa (PROVIDE). This interdisciplinary research program brings together scientists in the fields of environmental policy, environmental technology, development economics and environmental system analysis[1].

The PROVIDE project aims at developing socio-technical infrastructures which are sustainable both in environmental and social respects in applying Modernised Mixtures Approach (MMA). The PROVIDE program focuses on and contributes to the improvement of urban sanitation and solid waste management in East Africa (Kenya, Uganda, and Tanzania), with an emphasis on the Lake Victoria Region and capital cities. The project seeks to identify and assess viable options for improving the sanitation and solid waste situation in East Africa and contribution for realizing the millennium development goals (MDG). In the context of this interdisciplinary research programme, the present study seeks to highlight the role of households in the production and management of domestic solid wastes in particular.

1.3 Problem specification

In Dar es Salaam city it is estimated that over 60% of the population lives in informal settlements (DCC, 2004; Kyessi, 2002; Kyessi and Mwakalinga 2009; Sawio, 2008). The informal settlements generally are densely populated and lack good infrastructure, basic social services and amenities. The result is that a substantial part of the households in informal settlements is left without access to solid waste management making them particularly vulnerable. Most of the empirical data gathered in the present study are obtained from research conducted in Kinondoni, the largest and fastest growing municipality which covers a wide range of informal settlements in Dar es Salaam city.

---

[1] The project is coordinated by the Wageningen University and Research Centre, Department of Social Sciences, Environmental Policy Group (ENP). It is funded by the Interdisciplinary Research and Education Fund (INREF) of Wageningen University. The East African partners in the research program are: Ardhi University (Tanzania), Makerere University (Uganda) and Kenyatta University (Kenya). For more information see http://www.provideafrica.org.

Solid waste management at household level is inadequate and characterized by inefficient collection methods. In average, only about 40% of the waste generated is collected and deposited off (Chinamo, 2003), while a significant portion of households do not have access to proper solid waste services. The result is that household solid wastes are dumped in open pits, near houses, or in streets and in storm drainage channels. This creates various hazards of which the most serious is health risk, but other includes the existence of piles of rubbish, and the general lack of aesthetic appearance of the premises. Although solid waste collection is outsourced in most parts of Dar es Salaam, solid waste contractors (including community-based organisations (CBO's) and private companies) engaged in solid waste collection lack resources in terms of equipment, finance and technical expertise in solid waste and information management.

In addition, solid waste management is hampered by a lack of data at all levels from the household, the ward, and up to the municipal level. If data are available at all, they are generally unreliable, scattered and unorganized (World Bank, 2002a). As a result, the planning and implementation of solid waste management policies have remained a difficult task to accomplish. Although legislative and regulatory frameworks exist with regard to waste management, municipal authorities find themselves in situations where they are unable to collect and dispose the wastes generated by households The ever increasing amounts of domestic wastes are simply beyond their capacity for managing it.

It is against this background that the current study attempts to analyse the role of households as waste generators, waste handlers and service recipients in the context of Solid Waste Management practices in East African capital cities.

1.4 Relevancy of the study

The main relevancy of this study is to contribute to the achievement of the goals and targets of Millennium Development Goals (MDGs) by increasing access to solid waste management for households living in informal settlements in East African capital cities. The MDGs sharpened focus on many issues which can improve the lives of those who manage to live on incomes that is lower than the average. There are eight broad goals, with 18 specific targets and 48 indicators, and together they give direction to the collective commitment to progress in improving lives and living conditions. The goals are concerned with poverty, hunger, health, education, equal opportunities, the environment and international partnerships. Solid waste management can make a significant impact in achieving the Millennium Development Goals (CWG-WASH, 2006; Linares, 2003). The 10th and 11th targets under Goal 7 which aims at ensuring environmental sustainability and two of its targets are particularly relevant for the PROVIDE project:

- Target 10: halve by 2015, the proportion of people without sustainable access to safe drinking water and basic sanitation.
- Target 11: have achieved by 2020 a significant improvement in the lives of at least 100 million slum dwellers.

In an aim to contribute to the realization of the MDGs, this study intends to highlight and analyze in some depth the challenges facing contemporary solid waste management at household level in informal settlements in capital cities of East Africa.

The PhD-research contributes to the literature on household waste management by opening up the black box as it exists at the household level. What kind of waste handling practices (generation, storage, separations, collection, transport, re-use etc.) do exist inside the black boxes of households in informal settlements in East Africa? What kinds of formal and informal systems of service provisioning do exist at the household level? Which social and technical aspects of solid waste practices can help explain the so called lock-in effects which prevent the emergence of more effective and sustainable forms of management of domestic solid wastes in East African capital cities? By looking for answers to these general questions, this studies has a direct relevance for the MDGs.

1.5 Outline of the thesis

This study is organized into nine chapters, including this introductory chapter. The next chapter (Chapter 2) provides an overview of SWM in Dar es Salaam city, focusing on the background information of solid waste management in the city in general. It explains the state of solid waste management before and after privatization interventions, and highlights existing solid waste management practices. The chapter also describes the policy, institutional and regulatory frameworks within which the solid waste management actors operate, and it sketches the financial inputs to solid waste management as well. Chapter 3 deals with the development of a conceptual framework that guides the research and that forms the important contribution of the thesis. The research questions are presented based on the conceptual framework. The theory of Modernized Mixtures Approach is used as starting point to develop a conceptual framework for analysing the roles of households in SWM chains.

Chapter 4 presents a methodology and data collection techniques. The empirical results of the research are discussed in Chapters 5, 6 and 7. Chapter 5 analyses the role of households as waste generators in SWM chain by looking at the characteristics of the domestic wastes as generated by households. Specifically this chapter identifies per capita daily waste generation, the percentage waste fractions in household waste composition. In addition, the chapter identifies the factors influencing per capita daily waste generation and waste composition. Chapter 6 discusses the solid waste handling practices by households. The chapter describes and analyzes the waste management practices existing at household level as well as the solid waste flows travelling from the households to the waste transfer station. The chapter ends with discussing the roles of different household members in solid waste management activities. Chapter 7 explores the management of household wastes. It looks at the formal and informal stakeholders and their networks, it describes the relationships which exist between households and stakeholders, and the municipal assistance given to households in the context of solid waste management. This chapter likewise discusses and evaluates household's perceptions towards solid waste management services delivery. Chapter 8 explores the households and domestic waste management in Kenya/Nairobi and Uganda/Kampala in comparative perspectives. The chapter compares some findings from Dar es Salaam, Nairobi and Kampala on a selected number of SWM-aspects. Chapter 9, finally, summarizes the main findings of the study and presents the conclusions that can be drawn from the empirical chapters. It gives a reflection on the empirical results with the help of the conceptual framework used in this thesis. The chapter also highlights the limitations of the study, and recommends area for future research.

Chapter 2.
An overview of solid waste management in Dar es Salaam city

2.1 Introduction

Solid waste management is an increasing burden of the day for almost all urban centres in Tanzania, being most felt in Dar es Salaam city. According to UN-HABITAT (2004a,b) solid waste is one of the major environmental problems identified by Dar es Salaam City Council (DCC), and the Sustainable Dar es Salaam Project (1992), which is in need of immediate attention.

Since independence 1961 until 1994, Dar es Salaam City Council (DCC) was the sole provider of SWM activities in DSM city and it did not have any mechanism of involving the city residents in its undertakings (DCC, 2004). From 1994, the Dar es Salaam City Council decided to privatize some of its principal services in waste management, specifically waste collection and disposal. The privatization entailed the increased involvement of private and popular sectors and communities in solid waste collection. At the time, Dar es Salaam City Council failed to provide adequate solid waste collection services for the fast growing population due to inadequate resources in terms of finance, equipment, skilled personnel and lack of proper disposal sites (DCC, 2004). This state of affairs made the city authorities realize that it needed other partners particularly, private and popular sectors, in its efforts to build a sustainable city.

The rationale for this chapter is to provide background information of solid waste management in Dar es Salaam city in general in order to contribute to a better understanding of the state of solid waste management before and after privatization intervention, and the changing nature of solid waste management practices. The state of solid waste management before privatization is explained for the period between 1982 to 1994, while the situation after privatization is dated from 1994 to 2007. The review also aims to describe the policy, institutional and regulatory frameworks within which the solid waste management operates, and the financial inputs into the sector of solid waste management. This descriptive analysis with serve as background and supplement for the empirical findings reported in chapters 5 to 7 of the thesis.

The chapter is based on a study of various documents related to the question of solid waste management in the city and on discussions and interviews conducted with officials from National Environmental Management Council (NEMC), Ministry of Health and Social Welfare, City health officer and Kinondoni Municipal health officer in Dar es Salaam. The site seeing also provided the researcher with the actual situation of solid waste management practices.

Section 2.2 briefly gives the description of Dar es Salaam city. Section 2.3 presents the description of solid waste management system before and after privatization. Section 2.4 describes the policy, regulatory and institutional frameworks on solid waste management service delivery. Section 2.5 describes the solid waste management practices. Section 2.6 presents the waste stream for SWM in Dar es Salaam, while section 2.7 explains the solid waste management financial inputs and section 2.8 concludes the chapter.

2.2 Description of Dar es Salaam city

Dar es Salaam City is located in the Eastern part of Tanzania, a coastal city situated along the Indian Ocean. Dar es Salaam City is the main industrial and commercial centre for the United Republic of Tanzania. Dar es Salaam is one city, with 4 local governments: Dar es Salaam City Council and 3 Municipal councils of Kinondoni, Ilala and Temeke (Figure 2.1) and it covers an area of 1,800 km² both land and water (Chinamo, 2003).

According to the national census of 2002; the city had a population of 2.5 million with average growth rate of about 4.3% per annum with average household size of 4.2 (NBS, 2002). In 2007 it was estimated that its population had risen to about 3.0 million. High population growth rate is caused by urbanization process.

The urbanization process lead to ever-increasing haphazard and unplanned urban growth coupled with inadequate living standards, and severe deterioration of services and the city environment. Although there is a strategy aiming to achieve a city without slums' by 2015, some 70% of the population currently live in unplanned and un-serviced settlements (DCC, 2004; Sawio, 2008). The urban sprawl in the city as a result of this unplanned growth increases the burden of providing infrastructure services such as solid waste management services. According to Table 2.1, the population continues to grow fast, which is an indication of an increasing amount of solid waste generated as well. For example there was an increase in solid waste generation rate from 2,400 tons per day in 2002 to 3,456 tons per day in 2007 (Mkwela and Banyani, 2008).

In 2000, city council operations were decentralized into three municipalities: Kinondoni, Ilala, and Temeke. The municipalities were given full policy and legislative implementation authority. Planning and administration were done in a consultative manner (Venkatachalam, 2009). Based

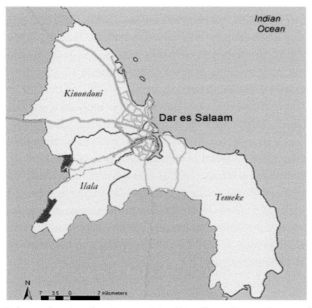

Figure 2.1. Dar es Salaam showing the city municipalities (http://www.dcc.go.tz).

Table 2.1. The population of the three municipalities of Dar es Salaam (DCC, 2004).

Municipality	2002	2003	2005[a]	2007[a]
Kinondoni	1,083,913	1,130,520	1,229,835	1,337,875
Ilala	634,924	662,225	720,401	783,687
Temeke	768,451	801,493	871,904	948,498
Dar es Salaam	2,487,288	2,594,238	2,822,140	3,070,060

[a] Projected population figures.

on the 2002 'Population and Housing Census', Kinondoni had the highest population, followed by Temeke and Ilala. Population figures of 2002 to 2007 are shown in Table 2.1.

Each municipal council provides public services such as health services, infrastructure including roads, community development, and solid waste management amongst others.

With respect to SWM the role of the Dar es Salaam City Council (DCC) and the three Municipalities is to regulate, supervise and control the activity (Mkwela and Banyani, 2008). The contracted organs (private companies and Community Based Organizations (CBOs) under DCC collect solid waste and dispose of it at specified dumpsites. Solid waste contractors provide service and charge the service recipients the amount set by the city and municipal authorities. During the period of undertaking this study the total number of solid waste contractors registered with the DSM City Council was 58, operating in Ilala, Kinondoni, and Temeke municipalities.

While this chapter gives an overview of SWM for the entire city, due to financial and time constraints the scope of empirical study was limited to Kinondoni municipality being the largest.

2.3 Description of the solid waste management system

The information given in this section is based on the interviews in October 2007 with Dar es salaam City health Officer, Principal Environmental Engineer from the Vice President's office, Kinondoni Municipal health officer, Ilala Municipal health officer, Temeke Municipal health officer complemented by information from documented literature on Dar es Salaam solid waste management.

2.3.1 The situation from 1982 to 1992

By the year 1982, the City of Dar es salaam, with an estimated population of 1 million was generating about 1,200 tons solid waste per day, while the capacity to collect and dispose of was only about 66 tons per day, or 5.5% of all refuse generated daily. The City then was dependent on 11 old and dilapidated trucks for haulage of refuse to a crude dumpsite, located in Tabata, a suburb to the west of the city centre.

Between 1983 and 1989 Dar es Salaam City received a grant aid from the government of Japan which made it possible to acquire new waste management equipment consisting of 30 refuse trucks, 3 skip-trucks, 30 refuse containers and 6 compactor trucks. The equipment raised the daily refuse collection capability to about 690 tons, out of the estimated generation of about 1,500 tons of refuse per day.

From 1990 to 1992 the situation had reverted back to its problematic situation when the running fleet was reduced to only 7 refuse trucks due to lack of funds for the maintenance of the equipment. By June 1992, 26 of the 30 tipper trucks, 5 of the 6-compactor trucks and 2 of the 3 skip-trucks were grounded. As a result, the City authorities were unable to provide adequate refuse collection services particularly in the city centre, and the percentage of the solid waste that was being collected each day dropped (Bakker *et al.*, 2000).

While the volume of solid wastes generated in 1992 amounted to 1,400 tons of waste a day, the DCC was only capable of collecting between 30 and 60 tons (2-4%) of this amount. The city environment was characterized by large amounts of dumped garbage in public open spaces, on streets and major roads and in open drains, resulting in flooded roads, ground water pollution, soil contamination, and escalating outbreaks of communicable diseases like cholera, diarrhoea and dysentery, flooding during rainy seasons, air pollution due to decaying of waste, and demolition of tourist interest and foreign business in the city. The situation was particularly serious in the central business district. The main reasons for the DCC's failure to manage solid waste was due to lack of equipment; lack of financial resources to purchase spare parts and fuel for the fleet, mixed signals on political will; un-focused City leadership and lack of an official disposal site (Halla and Majani, 1999). The failure of the solid waste management system caused alarm and concern, both at home and abroad.

2.3.2 Initiatives to improve solid waste management in Dar es Salaam city

In an effort to confront the worsening solid waste management services, the government approached different donors to assist in terms of equipment through Sustainable Dar es Salaam Project (SDP) financed by UNDP. The Sustainable Dar es Salaam city Project (SDP) was started in Dar es Salaam in January 1992 under the auspices of the Habitat Sustainable Cities Programme (UN-HABITAT, 2009; UNCHS, 1994). In 1993 the SDP introduced the Environmental Planning Management (EPM) process with the overall aim of supporting the Dar es Salaam City to promote Public Private Partnership (PPP). The main aim was to bring together the various stakeholders on the urban scene, including central and local governments, the private sector, various donor organizations, and the CBOs, to agree on strategies to address the environmental problems of cities. In August 1992, the SDP organized the Dar es Salaam City Consultation on Environmental Issues.

The SDP mandated stakeholders to produce a Strategic Urban Development Plan (SUDP) for the city. This was to be accomplished with substantive inputs from working groups (WGs) composed of 'informed' representatives of key stakeholders, including grassroots representatives. Their functions within local government were to increase the deliberative, consultative, strategic, and mobilization oriented strategies within the local administration (Kombe, 2001). Dar es Salaam City Council (DCC) representatives participated in the national and city consultation processes, which also included the Ministry of Lands and Human Settlements Development, the Ministry for Local Government and Administration, and national and international non-governmental organizations (NGOs). The purpose was to gather a range of inter-agency urban development stakeholders to discuss priority areas of intervention and develop a single response mechanism.

The City Consultation identified solid waste management as a priority environmental issue to be addressed immediately, recommending that cross-sectoral, multi-institutional working groups be established to implement a five-point strategy of intervention. This initiative included: launching of an emergency clean up campaign, privatization of refuse collection and disposal services, establishment of a community based collection system, promotion of waste recycling and composting, and better management of disposal sites (DCC, 1993; Kombe, 1995; Majani and Halla, 1999; UN-HABITAT, 2004b, 2009).

Launching of an emergency clean up

The emergence clean up campaign was tackled by a sub-working group named after this strategic element of intervention, with members from: DCC, SDP, PMO, donors such as Japan International Cooperation Agency JICA, private sector, Regional Administration, University of Dar es Salaam and the former Ardhi Institute, the UCLAS and now ARU commenced in 1992 with an intervention from the Prime Minister's Office (PMO). The campaign entailed removing accumulated wastes from market places, open spaces, major roads and streets. This was followed by the establishment of collection points in the city centre, market places and open spaces. The PMO with support from the Governments of Japan, Denmark, Italy and Canada raised US $ 1.4 million for procurement of solid waste collection equipment (bulldozer, a wheel loader, excavator, grader, etc.) and spare parts for repairing 30 garbage collection trucks, opening of a new dump site at Vingunguti and facilitating day to day refuse collection services. The emergency clean up campaign was quite successful, so that in a short period waste transported by the city council rose from the original 30-60 tons per day in 1992 to 300-400 tons per day in 1994. As soon as the above was accomplished the sub-working group was dissolved.

Privatization of solid waste management and dissolution of Dar es Salaam City Council

From 1994, the Dar es Salaam City Council decided to privatize some of its principal services in waste management, specifically waste collection. The privatization entails involvement of private and popular sectors and communities in solid waste collection.

It was estimated that, by 2002, the percentage of waste collected had risen to an average of over 30%, the collection being carried out by more than fifty enterprises, many of them small and based in the communities they served (Chinamo, 2003; Kassim, 2006; URT, 2003). The sub-working group on privatization of solid waste collection strategy members included DCC, SDP, contractors, Ministry of Industry and Trade, Ministry of Health, Ministry of Lands and Housing Development, National Income Generation Programme (NIGP), and Councillors. The strategy was demonstrated in ten wards[2] in the town centre, namely Kivukoni, Kisutu, Mchafukoge,

---

[2] A ward is a smaller administrative units of the municipality. The wards are constituted by a number of sub-wards areas (In Swahili a sub-ward is referred to as 'mtaa'. Each ward has a Ward Executive Officer (WEO) who represents the ward at municipal level. The ward executive Officer and the sub-ward or mtaa leaders are the government leaders in their local area. The ward Executive Officers are appointed by Government, while sub-ward leaders are elected by the residents living in their area. Each mtaa is divided into units of approximately 10 houses, these units are called ten-cell units, an organization of ten households.

Upanga East, Upanga West, Jangwani, Kariakoo, Gerezani, Mchikichini, and Ilala. Multinet Africa Company Limited won the contract to collect solid waste in ten out of 39 wards, making it the beginning of the privatization process of solid waste collection in Dar es Salaam City. The contractor started to operate in September 1994 using a few own vehicles and eight others hired from Dar es Salaam City Council.

The beginning was, however, difficult, because neither the city council nor the contractor had the experience of dealing with each other in solid waste collection, let alone the contractor's uphill task of collecting refuse collection charges (RCC) from the generators of the solid waste. The city council needed to facilitate the contractor in many ways, if the tasks at hand were to be done as expected. One such facilitation was to put in place a By-law (Dar es Salaam City Collection and Disposal of Refuse), which was passed in 1993 and took effect one year later in 1994. Moreover, city fathers had to enforce the by-law, and take to task all who do not pay their waste collection fees. And in collaboration with the contractor, the city council was responsible for creating public awareness on the privatization arrangement and roles of the various actors including the recipients of the solid waste collection services.

In 1995 the Prime Minister instructed the DCC to make the city clean within six months, due to unsanitary conditions in the city which were mainly due to uncollected solid waste. Several actions were taken including expansion of the privatization system to cover 25 wards, this being the only feasible options for DCC to pursue due to lack of resources. In May 1996, following revision of the contract, the tender for refuse collection from 25 wards in DSM including the 5 wards where Multinet was still operating took place. Two contractors were successful. However, the DCC made very little progress in cleaning the city for the given period of six months and DCC was judged to have failed in its mission with the Prime Minister (JICA, 1997). The failure of the city council to deal with refuse collection charges (RCC) defaulters, i.e. households and other service recipients refusing to pay the waste collection and disposal charges, amongst other things, contributed to the poor performance of the contractor whose collection of solid waste in the central area dropped from an initial 70% peak to 15%. Consequently, the privatization process was temporarily halted and downscaled to cover only 5 wards. Against this background the DCC-Council was dissolved by the Government in June 1996. The Dar es Salaam City Council was dissolved because it failed to comply with the provision of the local government (urban authorities) Act of 1982 and conducted its affairs in a manner that was incompatible with the realization of the purpose of that act. The Dar es Salaam City Commission (DCC-Commission) was therefore formed by the Prime Minister's Office to replace the Council. The DCC-Commission encouraged and speeded-up 'privatization' of SWM through Public-Private Partnership (PPP). When the newly appointed City commission started functioning in 1997, they found only 4 out of 30 trucks were in working condition (Bakker *et al.*, 2000).

The problems leading to that situation and the general experience were reviewed and became the building block for the start-up of phase two of privatization of solid waste collection services in 1996. In phase two 13 wards were added and four additional contractors won tenders for collecting solid waste in the designated wards. The contractors included: Mazingira Limited who were contracted to collect solid waste in Msasani, Kawe, Kinondoni, Mwananyamala, Manzese, and Tandale; Allyson's Traders in Magomeni, Mzimuni and Ndugumbi; Kamp Enterprises in Makurumla and Ubungo; Kimangele Entreprises in Keko and Temeke No 14. Solid waste collection in the remaining 29 wards at that time, continued to be served directly by the DCC. This approach had achieved considerable improvements whereby the rate of collection increased slightly to about 8.1% of the total waste generated, which was about 1,772 tons per day. This increase was a result of the emergency clean-up operation launched in 1992 under the guidance of the Prime Minister's Office. So in late 1998 preparations were made to extend this system to city's 52 wards (Burian, 2000; Kombe, 2001; Majani, 2000).

During phase two, the daily solid-waste collection increased in the newly contracted wards. Solid-waste heaps were reduced, especially in open spaces and market places. However, the constraints were similar to phase one, including inadequate payment of RCCs to the contractors. Preparations were insufficient to involve and raise awareness of people on the new strategies to clean the city and the responsibilities of individuals and stakeholders. Inadequate revenue collection prevented contractors from meeting financial targets. Contractors' equipment and facilities were inadequate, and they failed to meet promises to purchase replacements. DCC was unable to provide an enabling environment for the contractors (e.g. information on residents liable to pay RCCs, an effective public awareness campaign). The contractors required close supervision, monitoring, support for planning, technical advice and financial assistance.

All households were not treated equally in all wards. As noted by Bakker *et al.* (2000), according to the municipal by-law, RCC's depend on income category of the household. The low income households (informal settlements or unplanned areas) are charged 500 TZS ($ 0.65) per household per month, middle income households is 1000 TZS ($ 1.25) per household per month and high income households (fully planned) are charged 5,000 TZS ($ 2.5) per household per month[3]. As such, the municipal by-laws favour formal waste collection to planned settlements through privatization. DCC decided to treat households differently because the ability and willingness of the people to pay and the cost of the services were some of the factors used to determine the rates to be charged.

Establishment of a community based collection system

City authorities decided to consider involving community in SWM by supporting and promoting the establishment of Community Based Organizations (CBOs) that were interested in participating in solid waste collection activities. The sub-working group on Community Involvement in Solid Waste Management had representatives from communities, CBOs, non-governmental organizations (NGOs), Dar es Salaam City Council, Sustainable Dar es Salaam Project (SDP), National Income Generation Programme (NIGP), International Labour Organization (ILO, Vice-president's Office (Environment), and Ministry of Community Development and Children. Among other activities, this working group promoted establishment of CBOs interested in participating in solid waste collection activities. Examples of groups formed as a result of this initiative include Kinondoni Moscow Development Association (KIMODA) and Hananasifu Women Development Association (KIWODET). Due to narrow roads these CBOs collected solid waste in their respective informal settlements using push-carts and brought them to a transfer point where solid waste collection trucks collected and transferred them to the dumpsite. The CBOs were also engaged

---

[3] According to $ to TZS exchange rate in 1998.

in cleaning the storm drains, street sweeping, grass cutting and public awareness creation in their respective neighbourhoods[4].

Several companies from abroad approached the city authority with the intention of providing exclusive waste management service to the entire city using heavy machinery. The city authority rejected these offers, because they could not afford the costs involved and considered it too risk to grant one company a monopoly. The idea was also rejected because it meant that most of the residents employed in this sector would be rendered jobless. Which is why the city authorities decided to consider the involvement of other participants such as CBOs, NGOs and micro-enterprises. Many of these were based in informal/unplanned areas where the large enterprises were not willing to provide service, due to lack of necessary infrastructure. CBOs were considered more suitable to serve these areas, amongst others because of their close relation with the households and the community in informal settlements (Bakker *et al.*, 2000).

Promoting recycling and composting

Solid waste recycling was also incorporated in the waste management strategy. The sub-working group on solid waste recycling and composting had representatives from DCC, NIGP, community representatives, research institutions, SDP, CBOs and NGOs, Centre for Cleaner Production, and donors. Among other activities the subgroup encouraged the communities, CBOs and NGOs to promote recycling and composting by sorting solid waste at the source. The objective of promoting recycling was to reduce the waste generation and the need for dumping space and to create employment for the city residents.

The planned activities included the mobilization and support of small-scale recyclers and waste pickers (scavengers), awareness creation to industries to encourage waste recycling and product innovation, and market development for recycled products. Part of the latter was the promotion of the market linkage between industries and small-scale collectors, and exploring means to subsidize costs of transporting recyclables. Composting to support urban agriculture was considered, though most households were not aware of composting. The serious constraint regarding composting is that waste separation at household level is not practiced in Dar es Salaam.

Better management of disposal sites

Another issue tackled was an the development of an alternative disposal site. An interim dumpsite was identified at Vingunguti. With donor funding, the dumpsite was improved by the construction of a 1.2 km access tarmac road; installation of a weighbridge at the entrance of the dump site; and the construction and equipping of the dumpsite office with a computer for data recording.

Other improvements included enhancement of internal roads to facilitate smooth movement of trucks; control of leachate from the dumpsite, and the introduction of a refuse disposal charge

---

[4] Neighbourhoods of sub-wards are made up of several ten cell units. Ten cell leaders are like sub-ward (*mtaa*) leaders elected by the household members constituting the cell. Ten cell unit exist as the smallest political-administrative unit in Tanzania, in reality, however, the size varies between 8 and 15 houses.

to meet the operational costs of the dumpsite. The need for new disposal sites was also addressed by evaluating suitability of various potential sites within Dar es Salaam.

Institutional reform

Another step for improving solid waste management delivery for sustainability was to make an institutional change by forming separate departments dealing with solid waste which must be seen as an effort of decentralizing solid waste management services. In March 1998, the DCC established an independent Waste Management Department (WMD). The WMD was born out of the Dar es Salaam Sewerage and Sanitation Department (DSSD) and the units of the Health Department dealing with waste.

In 1999, the service was decentralized into three municipalities of Ilala, Kinondoni, and Temeke, with a central coordinating division at City level. The municipalities deal with operational activities of managing solid waste (Chinamo, 2003). The department of solid waste management is therefore run by the head of department, assisted by the officers in charge of the specialized functions of planning, monitoring and control, the coordination of municipal operations, vehicles and equipment supplies. The department's responsibilities are to support the planning and delivery of waste management services, development of clear policies and guidelines for the provision of services by the contractors.

Other responsibilities include: enforcement and compliance to health regulations, deployment and development of competent personnel, and improvement of solid waste management rules and regulations. Figures 2.2 and 2.3 show the arrangement of responsibilities at City and Municipality levels. The department of SWM in each municipality, under the full council and the municipal director, provide directions to the ward, sub-ward and the public concerning waste management.

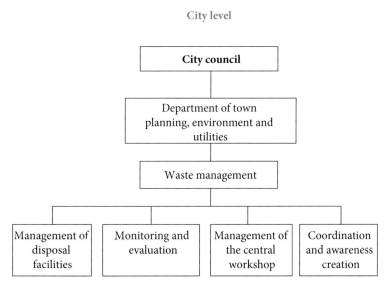

Figure 2.2. Arrangement of responsibilities at city level (Chinamo, 2003).

Municipal level

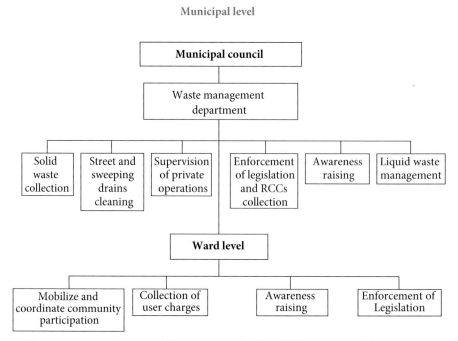

Figure 2.3. Arrangement of responsibilities at municipal level (Chinamo, 2003).

Each sub-ward is divided into units of approximately 10 households, these units are called ten-cell units. A ten-cell leader heads a ten-cell unit and one of their main task is to supervise the cleanliness of the units.

2.4 National policies and the regulatory and institutional frameworks relevant for solid waste management

This section describes the relevant national policies and the regulatory and institutional framework for solid waste management as applicable in Dar es Salaam and other urban authorities in Tanzania. As indicated by Kironde (1999), urban councils are charged with most day-to-day duties and responsibilities in dealing with urban waste. These duties and responsibilities are spelled out in a number of pieces of national legislation as outlined below.

2.4.1 Solid waste management policy framework in Dar es Salaam

In Tanzania, the solid waste management policy framework is embodied in number sector policies which provide guidelines in the sanitation and environmental management. In other words there is no single comprehensive document of solid waste management policy. The most important national policies containing sections on solid waste management and sanitation are the National Environmental Policy (NEP), the National Healthy Policy (NHP), and the National Human Settlements Development Policy (NHSDP).

National Environmental Policy

Tanzania National Environmental Policy (1997) emphasizes the promotion of safe water, environmental infrastructure to protect waste disposal services, the development of urban and rural waste management systems and the review of the laws governing hazardous waste. According to the NEP, the following are some reasons for deteriorating environmental condition: inadequate institutional coordination, inadequate monitoring and information systems, inadequate involvement of major stakeholders (e.g. local communities, non-governmental organizations and private sector) in addressing environmental problems, and inadequate financial and human resources.

In addressing environmental deterioration in urban areas, the policy sets out the following objectives: integrated planning, improved management of urban centres and designation of urban land uses to be based on environmental impact considerations, control of indiscriminate urban development particularly in vulnerable sites as coastal beaches, flood-prone and hilly areas and development of environmentally sound solid waste management systems (United Repbulic of Tanzania, 1997)

National Health Policy

The National Health Policy is aimed at providing direction towards improvement and sustainability of the health status of all the people, by reducing disability, morbidity and mortality, improving nutritional status and raising life expectancy. The policy recognizes that good health is a major resource essential for poverty eradication and economic development.

The objectives of the NHP include: (1) sensitize the community on common preventive health problems, and improve the capabilities at all levels of society to access and to analyse problems and design appropriate action through genuine community involvement; (2) promote awareness within the government and the community at large that health problems can only be adequately solved through multi-sectoral cooperation, involving such sectors as education, agriculture, water and sanitation, community development, women organizations, and non-governmental organizations; and (3) promote sound use of water, encourage basic hygiene practices in families, promote construction of latrines and their use in all households, health centres and educational institutions, and encourage the maintenance of clean environment around houses and institutions.

National Human Settlements Development Policy

The Ministry of Lands and Human Settlements Development (MLHSD) is the custodian of the National Human Settlement Development Policy 2000 (NHSDP). The objectives of NHSDP relevant to sanitation and solid waste management include: to protect the environment of human settlements and of ecosystems from pollution, degradation and destruction in order to attain sustainable development and to encourage development of housing areas that are functional, healthy, aesthetically pleasant and environmentally friendly. The policy also provides guidelines to the main actors (the Ministry responsible for Land and Human Settlements, local authorities and the non-state actors (private sector) so as to maintain environmentally friendly human settlements.

The local governments in Tanzania are the main implementers of the national policies[5]. Since the local councils are decentralised in this case, therefore, the implementation of the national environment policy is within the decentralized system. A study by Mruma (2005) observed that many implementers of policy are not aware of policy standards. The city lacked human resources and instruments for implementing environmental activities. Conflict between central government institutions responsible for policy formulation in Tanzania, which are National Environment Management Council (NEMC) and the Division of Environment over the roles in the management of environment also affects implementation of the environmental policy at the local level. In this case, environmental activities especially on solid waste management are not well coordinated with either the National Environmental Management Council or the Division of Environment. As we can note, these policies are very wide, covering many issues, and not easy to operationalise specific interventions. For example the policies do not state how households should participate in SWM issues and what areas require their involvement. In this case obviously households may not be aware of the policies. As noted by Mruma (2005) indicated that most of the local community are not aware of the environmental policies. Even most of the councilors[6] lack knowledge on the policy standards. Therefore, it becomes difficult to implement policy requirements.

2.4.2 The Tanzanian legislative and regulatory framework for solid waste management

This sub-section explains the legislative and regulatory framework for SWM in Tanzania and the main laws governing solid waste management in Dar es Salaam.

National Environment Management Council

The National Environment Management Council (NEMC) is under the vice-president's office. The main role of NEMC is to provide advice to the vice-president's office on all matters pertaining to environmental conservation and management. NEMC was established by an Act of Parliament No.19 of 1983 to perform an advisory role to the government on all matters relating to environment management. In addition it has the following roles: advise government on all technical matters for effective environmental management, co-ordinate the technical activities of all bodies concerned with environmental matters, enforce environmental regulations (norms, standards, guidelines and procedures), assess, monitor and evaluate all activities that have impact on the environment, promote and assist environmental information, communication and capacity building and seek advancement of scientific knowledge on the root causes of the changes in the environment and encourage the development of environmental sound technologies.

---

[5] The Central government Ministries in Tanzania formulate policies, delegate legislative powers, and grant resources to the local governments, which define the goals by formulating local environmental policies (by-laws).

[6] Elected local government representatives.

National Environmental Management Act

The Environmental Management Act 2004 (EMA), which became effective on 1st July 2005, is now the umbrella law on environmental management in Tanzania. The Act repealed the National Environment Management Council Act, 1983 while providing for continued existence of the National Environment Management Council (NEMC) under which it was established. The new EMA is comprehensive and has provisions for division of roles and mandates among actors involved in environmental activities including the Local Authorities.

Among the major purposes of EMA are to provide for legal and institutional framework for sustainable management of environment in Tanzania; to outline principles for management, impact and risk assessment, prevention and control of pollution, waste management, environmental quality standards, public participation, compliance and enforcement; to provide for implementation of international instruments on environment; to provide for implementation of the National Environmental Policy; to provide for establishment of the National Environmental Trust Fund and to provide for other related matters.

The functions of NEMC as per the EMA are mentioned in Sections 17 and 18, of the Act and include undertaking enforcement, compliance, review and monitoring of environmental impact assessments. The EMA No. 20 of 2004 prescribes the duty of local government to manage and minimize solid waste, disposal of solid waste from markets, business areas and institutions, storage of solid waste from industries and solid waste collection from in urban and rural areas:

- *Section 114* stipulates that 'it is the duty of the Local Government to manage and minimize solid waste';
- *Section 139* sets out powers of the Local Government authorities to minimize wastes; 'that each Local Government Authority shall have all powers necessary for purposes of preventing or minimizing wastes'.

Local Government (Urban Authorities) Act No. 8 of 1982

The Local Government (Urban Authorities) Act of 1982 gives considerable responsibility to urban authorities for waste collection and disposal. It requires urban authorities to, among other things, 'remove refuse and filth from any public or private place' (section 55 (*g*)). Also, urban authorities are required to provide and maintain public dustbins and other receptacles for the temporary deposit and collection of rubbish. Section 55 (*i*) provides for the prevention and abatement of public nuisances that may be injurious to public health or to good order. Urban authorities are also empowered to ensure that residents keep their premises and surroundings clean.

The Township Rules, made under the Township Ordinance of 1920

The oldest legislation dealing with SWM issues is the Sanitary Rules for the Township of Dar es Salaam made under the Township Ordinance of 1920, which gave the Medical Officer of Health powers to deal with sanitary nuisance and unsanitary premises. These rules have been retained over time and are operative under the Local Government (Urban Authorities) Act. To meet these responsibilities, the DCC drafted a number of by-laws relating to waste management. The most important of these are the Dar es Salaam Collection and Disposal of Refuse by-laws of 1993 dealing with collection and disposal of refuse based on the Local Government (Urban Authorities) Act no. 8 of 1982 under section 56 and 13. This by-law was passed in order to enable the privatization of waste collection and disposal. They require occupiers of premises to maintain receptacles to keep waste and bind the DCC to collect and dispose of waste. Among other things, these by-laws prohibits people from causing a nuisance and throwing or depositing waste on streets or in open spaces not designated as collection points.

The Township Rules impose the following requirements: (1) rule 23 requires the occupier of any building to provide a receptacle to store refuse. Receptacles must be maintained to the satisfaction of the city inspectors. Garbage bins should be placed alongside roads for collection; (2) rule 24 empowers the council to require a person to remove the accumulated refuse he or she deposits anywhere; (3) rule 27 requires the occupier of any plot or building to keep the surroundings free from accumulated refuse; and (4) rule 25 prohibits the throwing of refuse on any street or in any public area. Sanctions are a fine of up to 400 TZS or 4 months' imprisonment.

The Public Health (Sewerage and Drainage) Ordinance of 1955

Old as it is, this law gives local authorities (city/municipal) powers in dealing with public sewers, drainage, and latrines in new and existing buildings. Municipal Councils have made by-laws on solid waste management but the ordinance has remained the main law.

Town and Country Planning Ordinance Cap. 378

The Ministry of Lands and Human Settlements Development (MLHSD) is responsible for urban development, housing, land policy, land-use planning and land administration. It is responsible for preparing or approving land-use schemes, including those concerning land required for waste management (section 19 and 31 of the Town and Country Planning Ordinance Cap. 378 of 1956). Approval of dump sites is thus the responsibility of the MLHSD.

Dar es Salaam solid waste management by-laws

The local government Act of 1982 reveals that, to date, there is no policy for SWM at the national level; rather there are scattered pieces of legislation on SWM in different policies and city or municipal by-laws which are, for that matter, not supported by a principal law or policy on SWM. Owing to the state of affairs, the city and municipal authorities in the country handles solid waste management issues according to by-laws they set for themselves.

In regard to institutional arrangements, a major piece of legislation which guided Solid Waste Management in Dar es Salaam was the Dar es Salaam Collection and Disposal of Refuse by-law of 1993 made under sections 56 and 13 of the Local Government (Urban Authorities) Act No. 8 of 1982 (Nkya, 2004). This by-law was passed to enable the privatization of waste disposal and to introduce Refuse Collection Charges RCC[7]. As explained in section 2.1 of this chapter, before 1990, solid waste management in Dar es Salaam was a free public service provided by the Dar es Salaam City Council. Since 1992, the city's solid waste management system was reformed and contracted out to private sector operators. A private contractor is given a monopoly for delivering solid waste management services after a competitive pre-qualification, and is allowed to collect revenue through the refuse collection charges to cover their operational costs. After reforms in the city council, the four authorities; the city council, the Ilala, Temeke and Kinondoni municipalities, formed their departments on waste management (see section 2.3.2). Decentralization is observed in waste management activities whereby every municipality formulated its own waste management by-law. These are:

- Kinondoni Municipal Council Waste Management and Refuse Collection fees by-laws 2000 (Kinondoni Municipal Commission, 2000);
- Temeke Municipal Council (solid waste management) Collection and Disposal of Refuse by-law, of 2000 (Temeke Municipal Council, 2000); and
- Ilala Municipal Council Waste Management (collection and disposal of refuse) by-law (Ilala Municipal Council, 2001).

In these by-laws the obligations of residents (beneficiaries of SWM services) and service providers are prescribed as: occupiers of premises should maintain receptacles to keep waste, people are prohibited from causing a nuisance and throwing or depositing waste on streets or in open spaces not designated as collection points, beneficiaries are required to provide and maintain to the satisfaction of the DCC a receptacle for domestic refuse, of a sufficient size and fitted with good and effective lid, pronounces penalties (fines and/or imprisonment) for defaulters (Kinondoni Municipal Commission, 2000, 2001), and define where and how collection charges should be paid by the residents, with the respect of amounts for different generators. According to Nkya (2004) these regulations were not, however enforced fully uniformly.

2.4.3 The institutional arrangements for SWM in Dar es Salaam

Although waste management could be considered a local issue, the central government and national institutions play a big role and carry considerable responsibility in the whole system of urban waste management (Kironde, 1999). In Tanzania the central government and national institutions which play a big role and carry considerable responsibility in the whole system of urban waste management. The main institutions dealing with environmental issues including solid waste management and their responsibilities are: The Prime Minister's Office (PMO), and Ministry of Health and Social Welfare.

---

[7] Payment for solid waste collection and disposal services which solid waste contractors are supposed to collect from waste generators, including households as defined by the Municipal SWM by-laws

The Prime Minister's Office (PMO)

The Prime Minister's Office (PMO) is the overseer of all local authorities, Dar es Salaam City and its Municipal councils inclusive, through the Ministry of State for Local Government. It has a major influence on the performance of the DCC as it is responsible for approving the councils budget which includes SWM budget, council's by-laws and appoints senior personnel to the council. It also provides the mechanisms for legislative procedures for the council's activities. The major sources of revenue for the council are government grants, taxes and foreign aid. The PMO approves the allocation of funds from the treasury to local governments and handles any negotiations for external assistance.

Moreover, all by-laws made by local governments must be approved by the PMO, such as by-laws to keep the environment clean or to charge various levies. Therefore the efficiency of the relationship between the PMO and city and municipal authority has a major effect on the governance of sanitation and solid waste management. Furthermore, part III, section 13 (1) of the Environmental Management Act, 2004 states that the Minister responsible for environment shall be the overall authority in-charge of all matters relating to the environment and shall be in that respect responsible for the articulation of policy guidelines necessary for the promotion, protection and sustainable management of the environment in Tanzania.

The Ministry of Health and Social Welfare

The Ministry of Health and Social Welfare has overall responsibility for public health and special responsibility, through the Government Chemistry Laboratory Agency (GCLA)[8]. Ministry of Health and Social Welfare plays a key role in SWM, as they are responsible for public health which is linked to SWM. The ministry provides policy and guidelines on sanitation and solid waste management. Health officers who are responsible for the supervision of SWM service delivery are under the mandate of this ministry.

It will be clear that the central government has a major role in urban-waste management, chiefly at the level of policy formulation, but also at the operational level. Besides, in view of the unsatisfactory situation of waste management in Dar es Salaam, the central government has had sometimes to intervene directly to clean the city. The major intervention was the emergency clean up of the city carried out in 1993/1994, when the central government provided contingent resources to move tons of accumulated waste.

2.5 Existing solid waste management practices: some facts and figures

According to the interview with Dar es Salaam City health Officer, the current practice in Dar es Salaam with respect to solid waste management is to handle the waste wholesale. This is partly attributed to the fact that waste management authorities do not address individual sources of the waste, which are many, hence, difficult to deal with individually. Based on the interview it

---

[8] GCLA is an agency under Ministry of Health and Social Welfare responsible for registration and overall movement of industries and consumer chemicals.

can be concluded that, basically, management of solid waste in Dar es Salaam consists of three major components namely: solid waste generation and storage, solid waste collection and solid waste disposal.

2.5.1 Total waste generation in Dar es Salaam city

The Dar es Salaam city was estimated to generate about 3,456 tons of waste per day by the year 2007. This amount was produced in residential areas, commercial, institutions, market, industrial hospital. Table 2.2 below shows the amount of solid waste generates by each category. The amount of solid waste generated from households presents the largest amount followed by commercial and market waste. Waste arising from street sweepings is the least amount. Household solid waste was established to be 2,768 tons per day, on average of 0.4 kg/capita/day. It has been estimated that 75% of the household waste is organic in nature.

According to Table 2.3, Kinondoni municipality generates the highest amount of solid waste amongst the three municipalities; meanwhile Temeke municipality has the lowest amount of solid waste generated.

Table 2.2. Waste categories in Dar es Salaam.[a]

Waste category	Waste generation (tons/day)
Households	2,768
Commercial	128
Institutional	21
Market	140
Street sweepings	6
Others including hospital	393

[a] Interview with Dar es Salaam City Health Officer Mr. Chinamo on 29/08/2007.

Table 2.3. Solid waste generation in Dar es Salaam in 2007.[a]

Municipality	Waste generated (tons/day)
Kinondoni	2,206
Ilala	750
Temeke	500
Total	3,456

[a] Interview with Dar es Salaam City Health Officer Mr. Chinamo on 29/08/2007.

2.5.2 Solid waste storage

Waste storage is under the direct responsibility of the waste producer; in residential areas waste is stored in different types of containers, e.g. plastic bags, old plastic buckets, baskets, boxes, open piles, but invariably some people discharge waste without even any storage facility, indiscriminately dump waste in open spaces, storm water drains, valleys and along the roads. The DCC health officer affirmed that inadequate space for solid waste storage, inadequate storage facilities are the major constraints in solid waste management and lack of waste separation in particular in Dar es Salaam (Figure 2.4). Added to this he claimed that households and individuals are not committed and, therefore, hinder a more effective solid waste management.

Yet, the Collection and Disposal of Refuse By-laws, of 2001 Section 4 (1 and 2) and Section 5, requires all households to have two solid waste collection receptacles (one for organic and the other for non-organic waste) of not less than 40 l fitted with a lid. It has been noted that this by-law is not adhered to (Kaseva and Mbuligwe, 2005).

Figure 2.4. Existing solid waste storage facilities in Dar es Salaam (photo by the author, 2007).

2.5.3 Collection of solid wastes

Solid waste collection and disposal in Dar es Salaam city is carried out by the three municipalities of Kinondoni, Ilala and Temeke, and the solid waste contractors, while the city council is responsible for the disposal sites (Kaseva and Mbuligwe, 2005). The three municipalities are autonomous in operations. Solid waste is collected and dumped in landfills. Unfortunately only a fraction of the waste generated can efficiently be collected. In 2007 the DCC, reported that about 35% to 40% of the solid waste generated per day in Dar es Salaam city was managed (i.e. collected, stored, and transported to the landfill or recycled). The collection service is provided in residential, commercial, industrial and institutional premises. Table 2.4 shows the amount of waste generated and amount collected from 1994 to 2007.

Table 2.4. State of solid waste collection and disposal from 1994 to 2007.[a]

Year	Generation/day (tons)	Collection/day (tons)	Percentage
1994	1,500	185	12%
1995	1,620	230	14%
1996	1,772	260	14.5%
1997	1,850	300	16%
1998	1,980	376	19%
1999	2,144	450	21%
2000	2,200	352	16%
2001	2,300	483	21%
2002	2,400	720	30%
2003	2,600	780	30%
2004	3,091	849	27.5%
2005	3,156	900	28%
2006	3,350	1,207	36%
2007	3,456	1,382	40%

[a] Interview with DCC City Health Officer on 29/08 2007.

According to Table 2.4 above, the percentage of waste being collected from 1994 to 2007 has been increasing. This information provides insight into the impact of privatization of solid waste collection and disposal services, as well as the increase of solid waste generation as the population increases.

In terms of solid waste collection frequency, the collection from different sources varies. From households waste is collected between 2 to 3 days per week. Waste from business places is collected daily or on agreed number of days depending on the type of business; for example if it is restaurant waste it is collected daily. However, the DCC health officer reported that collection frequency depends on the condition of contractors waste trucks. It was noted that most of the waste trucks used by contractors are very old and poorly maintained. Also the solid waste labourers (crew members) do not use protective gears (Ishengoma, 2000).

According to the interview with Kinondoni Municipal health officer, in places where there is no solid waste contractor operating, municipalities provide a standby trailer to a ward or sub-ward and the responsibility of managing solid waste comes to resort under the Ward Executive Office (WEO) or sub-ward leader through the ward committee. During field observations conducted for the present study it could be observed that the ward committee organizes youth from the same ward to – under payment basis – collect wastes from residents and deliver it to the standby trailer. They practice house to house collection by using wheelbarrows and pushcarts because most paths and roads in unplanned settlements are inaccessible by vehicle. When the trailer is filled with waste, the municipality transports it to the dumpsite. The standby trailers are normally stationed in locations agreed upon by the WEO and the ward committee.

Solid waste collection in DSM can be divided into primary and secondary collection practices. Some CBOs specialize in primary collection, while private companies and some CBOs do both primary and secondary collection, taking care as well for transport of the wastes to the dumpsite (Ishengoma, 2000). The primary collection (from household to the waste collection points in the neighbourhood) consists of house to house waste collection. House to house collection is mainly done at the moment when the collection vehicles pass around the streets to collect garbage from households in order to transport it to the dumpsite. The collection system is more effective in planned areas since the houses are more accessible when compared to unplanned settlements.

Collection take the form of a 'bring system' when a vehicle stands at a certain agreed point where a number of households bring garbage to the vehicle (Figure 2.5). These vehicles have a system of horn or sound to alert households to bring out waste which is emptied directly into the truck. House to house collection is practiced whereby handcarts are used to dispose the waste at the contractor's collection point or at the municipal collection points.

Communal collection is also partly practiced, and households dispose waste in enclosures located along road sides (Figure 2.6). These places are located in places which are accessible by trucks. Majority do not prefer these communal collection points, because on several occasions waste at collection points tends to spread and litter the area around, and become illegal mini dumps

Figure 2.5. Primary collection of solid waste in Dar es Salaam (photos by the author, 2007).

Figure 2.6. Solid waste collection points in Dar es Salaam (photos by the author, 2007).

since anyone can dump waste of any kind which attracts scavengers, this eventually increases the cost of collection and transportation to contractors.

As a strategy of improving solid waste collection and transportation services in unplanned areas, the three municipalities of Dar es Salaam have recently procured standby trailers which have been provided to solid waste contractors (Figure 2.7). These are placed at points agreed upon by residents and solid waste contractor. It can be in an open space or any other place where contractors can easily manage the relevant equipment.

Figure 2.7. Standby trailers owned by municipalities in Dar es Salaam (photos by the author, 2007).

2.5.4 Disposal of waste

The most problematic functional element of solid waste management in Dar es Salaam has been identified as (the lack of proper) disposal methods and facilities (Kassenga, 1999). A manifestation of this problem is pollution of ground and surface water sources by leachate from poorly managed and illegal solid waste dumps. Minimizing waste generation by focusing on management practices at the source can save disposal sites space, reduce illegal dumping, and therefore, cut down on pollution potential from solid waste (Mbuligwe, 2002).

There has never been a designed sanitary landfill for disposing solid waste in an environmentally acceptable way in Dar es Salaam. Disposal of waste most of the time takes the form of crude dumping. Dar es Salaam City Council has been operating 4 dumpsites since Independence 1961 as shown in Table 2.5.

Table 2.5. Dumpsite in Dar es Salaam since 1961 (Kassim, 2006).

Dumpsite	Period of operation
Tabata	1961-1991
Vingunguti	1992-2001
Mtoni	2001-2007
Pugu	Prospect of serving for more than 30 years from 2007

Because of crude dumping Dar es Salaam City Council has been sent away by residents within the neighbourhood from dumping in places. The Tabata dump site in the city was closed following an August 1991 court ruling in favour of residents of the Tabata area who complained of air pollution caused by burning waste dumped at the site. Residents of Tabata area in the Ilala municipality had battled for two decades against DCC over using their land for waste dumping. The Vingunguti dumpsite caused conflicts between the Ilala municipal council and the community since it was not in the city master plan. The Mtoni dumpsite, which was 6 hectares in size, existed for only 6 years until it was closed in 2007. Mtoni residents protested against the dumpsite in their area saying it is a breeding ground for diseases such as cholera and dysentery (Kassim, 2006). The Dar es Salaam City Council (DCC) established a new dumping site at Pugu Kinyamwezi in Ilala Municipality, the site to which the Mtoni dumpsite was relocated from 1st of February 2007.

At the time of field work of this study the three municipalities and solid waste contractors were dumping waste at the Kigogo dumpsite which is located 8 km away from the city centre in the Kinondoni municipality. The dumpsite is located within a residential area and is operated on a temporary basis which can run for 5 years and was established after the neighbourhood community requested the municipal council of Kinondoni to reclaim land and save their houses affected by the landslide caused by Eli Niño rains. These requests were forwarded to the Kinondoni municipality since year 2000, the decision of the municipalities to open it in 2007 was reinforced by the shortcomings resulted from the official dumpsite which is located in Pugu[9].

However, the Dar es salaam City Council recognizes an official solid waste dumpsite located in Pugu in Ilala Municipality covering an area of 75 acres, 35 km away from the city centre. The dumpsite is owned, managed and operated by the city council, which charges a dumping fee ranging from TZS 1000 to 4,000 (US$ 1-4) depending on the tonnage (Interview with DCC health Officer).

2.5.5 Resource recovery

Resource recovery is the strategy to manage solid waste at a very low cost right from its source (household); this will help in reducing waste transportation costs. Recovery of resources from solid waste is achieved mainly through recycling, which mostly is practised by individuals. Studies on solid waste recycling and composting which have been carried-out in Dar es Salaam have recommended that recycling, reuse and composting are the most suitable approach to effectively manage solid waste in Dar es Salaam (Kassenga, 1999; Kassim, 2006). Kaseva and Gupta (1996) indicated that 14.7% of the total waste generated in the city is recyclable material which can be recovered from the waste stream, while JICA (1996) indicated that recyclable wastes constitute only 12% of the total waste amount. Kaseva *et al.* (2002) found that 50% of the recycled waste stems from the waste collected by both the municipalities and the private contractors, while another 50% is recovered by scavengers from uncollected waste. The recycled materials include paper, textile materials, metal, plastics and glass.

A study by Mbuligwe and Kassenga (2004) reported that recovery takes place to different extents at the source and disposal places, and applies mostly to household and commercial waste. The recycling activities take place at point of generation, collection, illegal dump sites and final

---

[9] Discussion with the Kinondoni Municipal Health Officer, 4/09/2007.

disposal sites. As indicated in Table 2.6 for household waste, on site recycling accounts for 114 tons/day, which is about 8% of total household waste generated. At the waste discharge place, recyclable items are scavenged prior to collection of the waste, for example at markets. The total amount of recycled waste at the discharge point is estimated to be 2.6 tons/day (see Table 2.7) for the whole city. During transportation of waste before subsequent disposal, recyclables may be taken out of the waste. In this case, sorting takes place as the waste is loaded into collection vehicles. The extent of recycling at this stage is very small, and this small component is incorporated in recycling at the final disposal point. Table 2.7 shows that illegal dumping places contribute 8.9 tons/day of recycled waste, while at the final disposal place about 2.1 tons/day of waste is recycled.

Composting is another way of resource recovery, however, it is not largely practiced in DSM. The study by Simon (2008) found that composting is practiced by some community based organizations involved in solid waste collection for their farms and sometimes for orders from farmers living outside the city. Compost is discouraged in residential houses due to smell and flies nuisance associated with it. As previously mentioned the other serious constraint is that solid waste separation at household level is not practiced in Dar es Salaam[10].

Table 2.6. Quantities (tons/day) of waste disposed of by different methods (Extracted from JICA, 1997).

Type of waste	Self-disposal	Discharge	Illegal dumping	Recycling	Total
Household	651.3	172.9	478.1	114.0	1,416.3
Commercial					
Restaurant	0	12.7	0	1.0	13.7
Guest house/hotel	0	1.6	0	0.2	1.6
Other	0	11.8	0	0	12.0
Institutional	2.1	8.6	0	0	10.7
Market	0	33.9	0	0	33.9
Street sweeping	0	1.3	0	0	1.3
Informal sector	0	56.5	226.2	0	282.7
Total	653.4	299.3	704.3	115.2	1,772.2

2.6 The waste stream and waste disposal practices

In Dar es Salaam city, the waste stream encompassing all the sources of waste constitutes: self-disposal, discharge, illegal dumping, recycling, collection, and final disposal (JICA, 1997; Mbuligwe and Kassenga, 2004). Table 2.7 presents the waste stream for various sources and methods of disposal in the city.

---

[10] Mixed household waste needs preliminary sorting before being composted to avoid risks of compost contamination.

Table 2.7. Summary of waste stream components (extracted from JICA, 1997 and Mbuligwe and Kassenga, 2004).

Waste stream component		Daily generation (tons/day)	
Component	**Sub-component**	**Amount**	**Total**
Generation	household	1,416.3	
	commercial	27.3	
	institutional	10.7	
	market	33.9	
	street sweeping	1.3	
	informal sector	282.7	1,772.2
Discharge			299.3
Collection			296.7
Self disposal	proper	392.9	
	improper	260.5	653.4
Illegal dumping	from generation	704.3	
	from collection	129.8	834.1
Final disposal			166.9
Recycling	from generation	115.2	
	from discharge	2.6	
	from illegal dumping	8.9	
	from final disposal	2.1	128.8

In the case of self-disposal, the waste generated by a source is disposed by the source itself within the source's property. Typical examples of self-disposal methods are burying of waste in pits and burning. Self-disposal may be classified as proper or improper, depending upon the location. For instance self-disposal by use of a pit within the premises of the waste generator is considered to be proper while for combustion of waste, this is proper in areas of low population density where the resulting public nuisance and pollution load is small. The converse applies to improper self-disposal. Table 2.7 shows that, 653.4 tons/day of the total waste generated is self-disposed.

Discharge means that the waste generated by a source is handed over to a waste collector or discharged at a certain place from where it is normally collected by another party. This component includes taking the waste to an approved collection point, placing it in a waste collection truck, dumping it at a waste collection point, paying a handcart operator to remove the waste, etc. Referring to Table 2.7 discharge waste amount is 299.3 tons/day. After discharge waste may either be collected or recycled. The waste is discharged by a source at a certain place when it is collected and transported to a final disposal site. As indicated in the table the amount of waste collected was estimated to be 296.7 tons/day.

Illegal dumping implies that the waste generated by a source is dumped outside the property of the source in an area where such behaviour is prohibited. The illegal dumping practice is very common, typically in close proximity to the source; e.g. Dumping at the roadside, in open spaces in drains, in the sea, valleys, etc. Table 2.7 shows that waste is illegally dumped by the generation source or after collection. As shown in the table, the illegal waste contributed by generators has been estimated to be 704.3 tons/day. The waste which is illegally dumped after collection is estimated to be 129.8 tons/day. Adding these figures, gives a total for illegal dumping of 834.1 tons/day. The illegal dumping of waste was anticipated to decrease with the increase in the collection amount by 2007.

Recycling means that the waste stream generated by a source, or part of it, is sold or given to an external agency, such as a person, shop, company etc. for reuse or recycling. For example, paper, or bottles may be sold or food and grass waste may be given away to people for animal feed. According to Table 2.7 recycling take place at 4 points, i.e. generation, discharge, illegal dumping and final disposal. The amount of waste recycled from all these points amounts to 128.8 tons/day.

As indicated by Mbuligwe and Kassenga (2004), for restaurants 93% of the waste is collected for final disposal, while 7% is recycled. For institutional waste, 80% is collected, while 20% is self-disposed of. Apart from trunk road waste, 100% of street sweeping waste is collected and disposed of at the dumpsite. For the informal sector waste, 80% of the waste is disposed of by illegal dumping, while the remainder is discharged for subsequent collection at such places as market waste collection points. Table 2.6 presents data on quantities of solid waste disposed of by different methods.

The waste stream was developed in the context of the JICA study to forecast volumes of different waste streams in order to establish a proper management system for solid wastes by the target year 2005. While the figures on the different waste volumes are out-dated, the waste stream gives insight in the overall management system for solid waste the authorities had in mind.

Major problems facing solid waste collection and transportation services in the city include inefficiency of the transportation system due to frequent vehicle breakdowns, inadequacy of collection vehicles, and inaccessibility of some waste sources, such as unplanned undeveloped areas, due to poor road conditions. Mbuligwe and Kassenga (2004) reported that these problems are aggravated by non-enforcement of relevant solid waste management by-laws and regulations by the DCC.

2.7 Resources available to the Dar es Salaam City Council for solid waste management

2.7.1 Financial resources

This section explains the financial and technical resources available to DCC in managing solid waste. Table 2.8 presents the financial resource in SWM from 1993 to 1995, and 2005 to 2007.

Table 2.8 shows few examples of the expenditure before SDP commenced in 1992 and after it ended in 1996. For example, the DDC spent an estimated 31 million TZS on solid-waste collection in 1987/1988 (Kironde, 1999). In 1993, the DDC's total budget for solid-waste management amounted to around 194 million TZS, and around 150 million TZS of this was provided through a special intervention from the PMO as part of the emergency clean up of Dar es Salaam. As explained in section 2.3, DCC got donor support in managing solid waste under the Sustainable Dar es Salaam Project (SDP). This explains high expenditure in SWM in 1993 as indicated in

Table 2.8. Dar es Salaam City Council financial resources.[1]

Year	Expenditure on solid waste management (TZS)[2]
1987/1988	31 million
1993	194 million
1994	44 million
1995	268,872.6 million
2005/2006	4,253 million
2006/2007	7,924 million

[1] Data from Kironde (1999) and an interview with Kinondoni health Officer on 4/09/2007.
[2] In 1998, 665.8 TZS = 1 USD.

Table 2.8. The Sustainable Dar es Salaam Project SDP began with a budget of USD 696,000 intended for a two-year period (nearly all from UNDP) for the period 1992-1996 (UN-HABITAT, 2004b).

In 2005/2006, the expenditure was 4,253 million, and 7,924 million in 2006/2007 respectively. DCC finance their assigned expenditures (including SWM) from three main sources: central government, own-sources and local revenues. According to Venkatachalam (2009) central government grants account for nearly 90% of DCC revenues, while own-sources comprise 10% of DCC revenues.

2.7.2 Technical resources

As shown in Table 2.9, since 1987 the DCC has procured 30 tipper, 3 container, and 6 compactor trucks as well as 3 bulldozers, but the number in operation has been dwindling. The trucks were allocated over three districts (Kinondoni, Ilala, and Temeke).

Table 2.9. Acquisition of trucks and number of operational trucks and other equipment for waste management in Dar es Salaam, 1987-1995 (JICA, 1997).

Year	Equipment procured	Number of operational equipment			
		Tipping truck	Skip truck	Compactor truck	Bulldozers
1987	30 tipper trucks, 3 skip trucks	30	3	0	0
1988		30	2	0	0
1991	6 compactor trucks	28	2	6	0
1992		28	1	2	0
1993		26	4	1	0
1994	3 bulldozers	24	1	1	3
1995		20	1	1	1

The DCC has had grossly inadequate equipment and money to purchase fuel and spare parts and to pay incentives to workers. Some of the vehicles have been unsuitable and have been plagued with spare-parts problems. Thus, although it could be argued that the vehicles, for example, were inadequate, the DCC was unable to operationalize even those few that it had. The DCC did not have enough resources to keep its fleet of trucks working. The fleet was working at less than 20% of its 1995 potential capacity, as a result of the shortage of funds for fuel, maintenance, and labour. Consequently, the DCC leased some of these vehicles to the private waste-collection contractor.

According to an interview with the Kinondoni Health Officer, Kinondoni municipality had 9 waste trucks, out of which 7 were provided by the government of Japan, and 27 movable standby trailers. Temeke municipality had 5 trucks and 9 collection points. Trailers are distributed to waste contractors providing services to informal settlements.

2.8 Conclusions

This chapter has provided an overview of solid waste management of Dar es Salaam city from 1982 to 2007, specifically looking at the service provisioning, the problems, the initiatives to improve the situation of solid waste management, the existing solid waste management practices and the institutional and policy frameworks for waste management. The chapter has revealed that the state and performance of the solid waste management system in Dar es Salaam city has been an on-going matter of serious concern, periodically renewed strategies for improving, sustained levels of underperforming, and subject of criticisms and protests from the side of the local population.

Until 1994 when privatization of solid waste collection commenced in Dar es Salaam, DCC was the only authority which had the responsibility of providing SWM services to the city of Dar es salaam. However, DCC failed to provide efficient and adequate services to the residents of DSM. This was due to lack of funds for operation and maintenance of equipment. The Sustainable Dar es Salaam Project (SDP), through the Environmental Planning and Management (EPM) authority formulated strategies to address the worsening situation of SWM. An agreement to adopt the EPM in Dar es Salaam was signed by the government of Tanzania in 1992. The EPM promoted the inclusion of different stakeholders to create partnership arrangements which were judged to be necessary in order to resolve the problems of SWM in the city in the future.

The strategies which were formulated by the EPM included: Emergency clean up campaign; Privatization of refuse collection; Community management of solid waste; Refuse recycling and composting; and Management of disposal sites. This experience demonstrated that DCC alone could not deal adequately with the increasing volumes of generated solid wastes in the city. In other words, it showed that tackling the SWM problems needed a shared vision and concerted actions. It was judged to be better for SWM when there would be no single stakeholder in charge, but instead SWM would involve different stakeholders and in particular those who are affected by the problems. The privatization process was judged to have at least temporarily contributed to the improvement of solid waste collection and disposal. For example there was an increase in the collection rate from 2-4% in 1992 to 40-45% in 2003.

This review has provided an understanding of the general solid waste management situation in Dar es Salaam before and after privatization while also depicting the institutional and legal frameworks within which the solid waste management system and its actors operated. It is important to note that all the initiatives to improve SWM concentrated on the solid waste management at the city and municipal levels, while focusing on all categories of solid waste presenting the different waste categories, wastes from household form by far the largest category of solid waste. Amazingly, however, the existing policy and institutional frameworks have paid no specific attention to the roles of households in solid waste management. In an effort to make up for this blind spot, the current study focuses on solid waste management from the perspectives of urban households in particular. Households and also CBOs and NGOs will be researched for their contribution to the overall waste management systems while emphasizing their particular roles, perceptions and responsibilities as they stem from and relate to practices of waste handling in the primary phase of the waste chains as distinct from the actors and dynamics in the secondary phase of the waste chains.

The next chapter will discuss the conceptual framework guiding this research and the methodology used in the study.

Chapter 3.
Understanding and improving household-waste management in East African capital cities: conceptual framework

3.1 Introduction

As explained in Chapter 1 of this thesis, urban solid waste management in East Africa in general, and in capital cities in particular, has been a serious environmental problem and challenge to the local authorities. In particular household waste has become a major problem. The conceptual framework of the Modernized Mixtures Approach (MMA) can be used to formulate research questions guiding the understanding of household waste management in this study. The household waste management practices and their relationships with solid waste infrastructures are at the core of the framework, which also incorporates a range of other variables with relevance for understanding domestic waste management practices.

The study objects derived from this conceptual framework can be specified within three main areas:

- Households as *waste generators* delivering their domestic wastes to waste management chains and infrastructures.
- Households as *waste handlers* operating in interdependence from other actors in solid waste management chains.
- Households as *stake holders and recipients* of solid waste management service provided by other chain actors.

The basic concept of Modernized Mixture Approach will be used to further develop the conceptual framework for studying and analysing household waste management in East African capital cities (Tanzania, Kenya and Uganda) with respect to all three main objectives.

Before elaborating on the MMA and its basic concepts, we will first introduce the Integrated Solid Waste Management (ISWM) framework, which can be considered as a predecessor of the MMA and regarded as a tool that helps to address the challenges of solid waste management. This is discussed in section 3.2. Section 3.3 explains the concept of Modernized Mixtures Approach in solid waste management. Section 3.4 discusses the concept of households, and describes households as (key) elements and stakeholders in solid waste management chains. Section 3.5 presents the developed conceptual framework, conceptualizing the roles of households in domestic solid waste management chains. The chapter ends with presenting research questions in section 3.6.

3.2 Integrated solid waste management

The failure of large-scale technical systems in a number of areas has stimulated searches for alternatives which do not demand long development times or large commitments of capital. These alternatives are inspired by the so called Integrated Solid Waste Management approach or model. A number of studies have introduced conceptual models for analysing integrated solid waste management (Schübeler *et al.*, 1996; Tchobanoglous *et al.*, 1993; UNEP, 2009; USEPA, 2002; Van de Klundert and Anschütz, 2000). These frameworks were proposed with the aim of assessing existing SWM systems and to plan for more effective approaches towards municipal waste[11]. The ISWM concept was established by the US Environmental Protection Agency (EPA) in the early 1990's to expand existing solid waste management initiatives. ISWM provides an integrated, holistic approach to tackling solid waste management problems that aims to avoid many of the failings of previous technology driven approaches (Van de Klundert and Anschütz, 2001). Instead of focusing only on the disposal of solid waste, ISWM includes preventing waste, minimizing the initial generation of materials through source reduction, reusing and recycling, and composting to reduce the volume of materials being sent to landfills or incineration. We briefly discuss the concept of ISWM as it was developed by WASTE[12] (Van de Klundert and Anschütz, 2000), because the Modernized Mixtures Approach can be argued to build upon and extend this approach.

The concept of ISWM discussed by Van de Klundert and Anschütz recognizes three important dimensions in waste management: (1) the stakeholders involved in waste management, (2) the (practical and technical) elements of the waste system and (3) the sustainability aspects of the local context that should be taken into account when assessing and planning a waste management system. The important principle of ISWM is that a waste management system should be appropriate to local conditions and be feasible from a 'technical, environmental, social, economic, financial, institutional and political perspective' (Anschütz *et al.*, 2004). ISWM differs from conventional approaches towards waste management by seeking stakeholder participation, covering waste prevention and resource recovery, and promoting an integration of different habitat scales (city, neighbourhood, household). ISWM can be used as an analytical tool for the analysis and assessment of the whole project cycle, especially for design, formulation, for monitoring and evaluation of a waste management project.

The concept of Integrated Sustainable Waste Management (ISWM) is an approach to reach better, more sustainable solutions to solid waste problems, especially in cities in the South. Africa being in Southern regions, ISWM projects has been implemented with no exception to East Africa. For example UNEP and ILO have assisted urban authorities to implement ISWM plans in capital cities of East Africa. In Dar es Salaam, the International Labour Organization (ILO) in collaboration with the UNCHS Habitat assisted the Dar es Salaam City Council in the implementation of an integrated solid waste management programme in 1998. The ILO asked WASTE to provide assistance in the field of community-based waste management and small-scale recycling. WASTE assisted in the development of a waste collection plan for two communities and presented options for the enhancement of the waste paper, plastic and sheet metal recycling sector. CBOs were trained in running their own waste collection businesses.

In Nairobi, UNEP is closely working with the Nairobi City Council (NCC) towards the development of an ISWM plan for Nairobi that will include all aspects from waste minimization, segregation, collection, transportation, reuse/recycle, resource recovery, treatment and disposal. Whereas, in Kampala, Local Agenda 21 programme of UN-HABITAT, the United Nations agency

---

[11] Municipal waste is defined to include waste from households, non-hazardous solid waste from industrial, commercial and institutional establishments (including hospitals).

[12] Advisers on Urban Environment and Development in located Gouda, the Netherlands.

for human settlements, initiated the development of an ISWM system for Kampala with a human rights-based approach. The ISWM is necessary because all the current projects of infrastructure, service delivery, local government development and private sector involvement work in isolation. Also, the present system does not incorporate the 3Rs (reduction, recycling, reuse) necessary for the process to become environmentally sustainable (Dastidar *et al.*, 2007).

MMA was established to build upon and expand the ISWM approach and initiatives. In East Africa MMA is relevant in particular in the context of cities with high population densities living in slums or informal settlements where there are inadequate solid waste management infrastructures available or even absent. As indicated by Oosterveer and Spaargaren (2010), MMA is helpful to specify an approach to urban infrastructures especially in these situations with fragmented or absent infrastructures as starting points. Because this is particularly the case in informal settlements and slum circumstances, MMA shows a strong focus on the (non) servicing of poor people, who are most directly affected by poor solid waste management services. Especially for these situations, the MMA suggest and promotes a modular approach, starting from the local (field) conditions both in social and technical respects. As discussed in Chapter 1, this study is applying MMA in the context of the project Partnership for Research on Viable Environmental Infrastructure Development in East Africa (PROVIDE), contributing to the MMA development from a specific emphasis on the role of households in solid waste management.

The following section will present and explain the concept of MMA as the overall background to the research. Subsequently, a conceptual framework will be developed to guide the research on the role of householders in SWM in particular.

3.3 The concept of Modernized Mixtures Approach in solid waste management

The Modernized Mixtures Approach is an integrated approach to (water and solid waste) infrastructure development, which is in particular relevant for situations of poor developed infrastructures and poor people. So in a way it can be regarded as a 'pro-poor' operation method, so that the people, who need the waste management most, are explicitly addressed. Following the MMA this goal can be reached best if systems are designed on local, small or medium scales (households and/or neighbourhood-scales), depending from the starting situation. 'Modernized mixtures' supports a demand driven and participatory way of developing infrastructures which matches with policy guideline about involving users and stakeholders in sanitation and solid waste management in decision-making processes.

MMA is an approach which deviates both from the well-known western large scale high technological grid based systems, as well as from familiar small scale, low-tech, decentralized and stand-alone technologies that are currently being applied in many African communities. It integrates the (Working Group of the Millenium Ecosystem Assessment) technological, economic, social and governance dimensions of new environmental infrastructures against the background of specific local contexts.

Applying such a MMA to sustainable urban development means the introduction of an 'organized eclecticism' by combining various levels of scale, strategies, technologies, payment systems and decision-making structures, to create a better fit with the physical and human systems for which they are designed. This approach is referred to as 'mixture' because it takes the best features out of both (modern) decentralized and centralized systems, and combines them into hybrid solutions which better fit the local situation (Spaargaren *et al.*, 2005). The concept of Modernized Mixtures in this study refers to the development of environmental infrastructural systems which build upon and are constructed from decentralized units of the decentralized systems type and which take into account the specific local conditions of study areas in East Africa in both social and technological respects. East Africa is one of the Sub-Saharan African regions where the slum population is expected to grow from 101 million in 1990 to 313 million in 2015 (Oosterveer and Spaargaren, 2010). Given the problems of urban environmental infrastructures provision in slum areas and great environmental pressure, there is a need to reconsider measures that are feasible and applicable in East African urban centres and offer a certain degree of choice about the types and levels of (waste) services, and about the best way to combine large and small, high and low-tech systems into a single system.

In MMA optimization of technologies and technological systems for sanitation and waste can take place along all three dimensions as suggested by Van Koppen (2004), where the different designs can be optimized along the lines of flows, institutions and economics. These improvements have to be judged/assessed with the use of three sets of criteria: (1) ecological sustainability, (2) accessibility, and (3) flexibility and resilience in both technological and institutional respect. The ecological sustainability criteria are used to assess to what extent the new systems or the new technological options that become part of the emerging infrastructural networks or infrastructures improve the environmental performance of the urban (waste) infrastructure. The accessibility criteria are used to assess to what extent specific groups are included or excluded from the environmental infrastructures due to financial, physical or cultural reasons. These criteria are in particular relevant to make judgements about the accessibility of new infrastructures and services for poor people: the so called 'pro-poor infrastructures'. The criteria of flexibility and resilience (in both technological and institutional respect) are used to assess how easily or difficultly the system or unit can be build into more embracing future systems and how the (sub) systems behave in times of instability in various dimensions (climate, political, economic, and institutional). In contrast to large-scale technical systems, Modernized Mixtures finally do not demand such long development times or such large commitments of capital as centralized systems do.

Figures 3.1 and 3.2 together illustrate the basic notion of the modernized mixtures approach, bringing together elements from both the central and the decentral paradigm while combining them into a number of options and strategies which can be said to be better adapted to the particular local circumstances both in technical and social respects. Figure 3.1 represents the relevant dimensions or variables which have to be taken into account when developing urban environmental infrastructures for water and waste (water) services, while Figure 3.2 illustrates some possible ways in which these dimensions can be combined into specific modernized mixtures.

By moving towards the upper-right corner in the model centralized infrastructures tend to resemble the large scale public, central grid-based systems in industrialized, developed countries. Moving to the bottom-left corner in this model decentralized visualizes the decentralized and small-scale systems developed in the past for developing countries and particular Decentralized Sanitation and Re-use (DeSaR) solutions for industrialized countries. Literature has described DeSaR systems to be developed partly in opposition to centralized systems and claim to be more robust and cheaper in comparison to centralized systems and to deal more effectively with

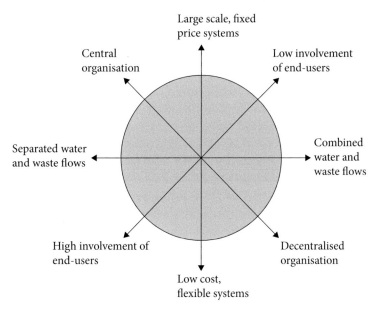

Figure 3.1. Dimensions of environmental infrastructures (Spaargaren et al., 2005).

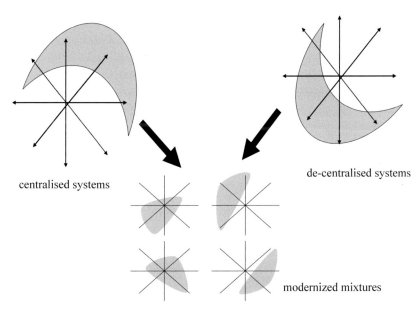

Figure 3.2. Modernized Mixtures Approach as alternative to central and decentralized systems (Spaargaren et al., 2005).

environmental challenges like high levels of water consumption and the indiscriminate discharge of potentially valuable substances from wastewater (Spaargaren *et al.*, 2005).

The third example of systems labelled as Modernized Mixtures shows different configurations which are adapted to the specific local context and requirements. Integrating knowledge with respect to all relevant dimensions as mentioned in Figure 3.1 is needed in order to optimize the chances for socio-technical systems to fit into the specific local social and technical conditions.

This new, hybrid paradigm or MMA can be characterized by its multidimensional character in both technological (scale, process and combination of water and waste flows) and management/governance respects (involvement of end-users, financial arrangements and organizational set-up). So similar to the ISWM model, also the MMA opts for an integrated approach, including all steps in the urban solid waste and the sanitation chains, combining multiple scales in organization, management and governance, and requiring the inclusion of technical as well as social scientific knowledge. The MMA shows particular sensitivity to making infrastructures fit the local situation, also and in particular so when the starting situation includes poor and low income users, fragmented infrastructures and difficult socio-spatial conditions.

The objective of the MMA is to create a 'fit' between different infrastructural options on the one hand and the prevailing socio-economic, ecological, technological, and political conditions on the other. For realizing this objective one has to develop and built upon a profound understanding of the specific (urban or semi-urban) settings in which these infrastructures are to be realized. This understanding will allow the involved stakeholders to answer the question, which technological options (and combinations) are realistically possible in the context of their particular cities. This means that each city, or even each neighbourhood, will require a specific mixture of technologies and institutional arrangements. Hence, working with the MMA means using a flexible, modular approach to urban environmental facilities and departing from a one-size-fits-all solution.

The general methodology of the MMA as outlined in this section stood at the basis of the conceptual framework for studying and analysing household waste management in Dar es Salaam city. This conceptual framework is inspired by the MMA but is more specific because of its particular focus on the roles of household in handling solid wastes. Before presenting the conceptual model, we first discuss the concept of household in more detail.

3.4 The concept of households

In this study the household is regarded as the most central unit of analysis. At the household level, solid wastes are generated and handled in ways that (co)determine the fate of the wastes later on and higher up in the waste chains. But how should the household be defined in the first place? management because it is the place where solid waste is generated. A number of authors have described the concept of household in different ways, relating to different social contexts. According to Oosterveer and Spaargaren (2010), the household is important because it is the place where people live and sometimes also work together, producing, storing, separating and re-using domestic wastes. Also the main end-users of solid waste management services are located in households. Niehof (2004a,b) describes households as the locus of livelihood generation, where resources are generated, organized, managed and used for economic activities as well as for the welfare of household members and care.

As indicated by Mtshali (2002), the concepts of family and household are often regarded as interchangeable, but they are not. The household is the most important institution in which people live. It is a basic unit of society where individuals both cooperate and compete for resources. The UN (2004) has defined a household as either: (a) a one-person household, defined as an arrangement in which one person makes provision for his or her food or other essentials for living without combining with any other person to form part of a multi-person household, or (b) a multi-person household, defined as a group of two or more persons living together who make common provision for food or other essentials for living. This definition is somewhat Eurocentric, lending itself well to representing stable nuclear conjugal family units, but more problematic when forming the basis for data collection in sub-Saharan African countries where prevailing models of social organization may be much more diverse and flexible (Van de Walle, 2006).

Dharmawan (1999) describes, 'A household as an organisation of human beings living in a common residence that disposes resources and pools incomes of its members and uses it by way of productive and reproductive activities for ensuring and securing its members' existence, in which socio-economic relations that are internal and external to the household unit, are continually formed for enabling in meeting such a need. Nombo (2007) has described households as the unit into which livelihood generation is anchored. It is the arena where much of daily life takes place and the centre of processes that determine the welfare and wellbeing of the individual members. The household is the context in which members interact and pursue the activities to provide for their daily needs and well-being.

The definition used by the National Bureau of Statistics (NBS) of Tanzania in 2002 census is more practical, but nevertheless quite comparable with the definition used by Dharmawan. This bureau (NBS 2002) defined households as: 'A one-person household is a household where a person lives alone in a whole or part of a housing unit and has an independent consumption. Multi-person household is a household where a group of two or more persons occupy the whole or part of a housing unit and share their consumption. Usually, households of this type contain a husband, wife and children. Other relatives, boarders, visitors and other persons are included as members of the household if they pool their resources and share their consumption'.

Building on the above definitions, the definition of household is therefore given to correspond with the households targeted in this research. Household in selected research areas consist of both family and non-family members that share (a part of) a housing unit and their resources, whereas, household members pursue economic and social activities within household premises to provide for their daily needs and well-being. This definition is considered to be relevant to this study because economic and social activities which are conducted by households within their premises lead to generation of solid wastes. Therefore the proper management of domestic solid wastes within the household is important to avoid serious health and environmental problems. Within the household, members of the household have different roles to perform in different (also waste-handling) activities, based on gender primarily. This definition of households (including economic/productive activities into the 'social' definition of the household as a family) is used in the present study to observe and analyse the roles of household members in SWM activities in Dar es Salaam.

In Dar es Salaam (NBS, 2002) households were grouped according to income and the marital status of the head of the households. According to distribution of expenditures, households are categorized as: 'low' – 50-100 USD/month, 'medium' – 100-150 USD/month, 'high' – 150-200 USD/month, and 'very high' – more than 200 USD/month. Marital status of the head of the household are categorized as male-headed, female headed or children headed. This categorization depends on who is the breadwinner of the household. NBS (2002), reported that women headed households constitute 21%, while households headed by men were 79%.

3.4.1 Households as (key) elements and stakeholders in solid waste management chains

This section presents the conceptual framework of the study, while elaborating on the core concepts contained in the framework. The conceptual framework allows for a specification of the research questions that are central to the PhD-research. These research questions will be presented after the discussion on the conceptual framework. In particular, we elaborate in this section how households can be assigned a distinctive position within the analysis of SWM chains compared with the other stakeholders and how to study their mutual relationships and modes of interaction with other actors in the chain.

In solid waste management, households take up a particular position in the flow of urban solid waste as they are to be considered the most central actors in the so called primary phase of the solid waste collection system or chain (Figure 3.3).

Households are the main producers of solid waste and the first responsible actors for dealing with wastes in the so called 'primary phase' of the collection-transport-disposal process of flows of domestic urban solid wastes. It is within households that primary collection and short term storage of the domestic waste is organised while householders are responsible for linking and coordinating their waste activities with the activities of the (system) actors operating in the secondary phase of

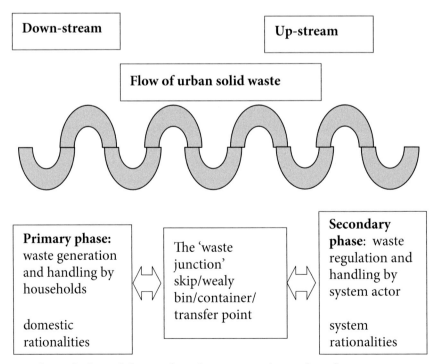

Figure 3.3. Flows of urban solid waste through primary and secondary phase.

collection, transport and disposal. An example of such a coordination is the adaption of the time schedules of household to the collection schemes of waste collectors.

Improving collection systems or developing waste management programs such as recycling and reuse systems require the involvement of households as they have to adapt and change their existing practices. In this respect it is important to be aware of the fact that the organisation of household activities are based on particular *domestic rationalities* and not necessarily on the effectiveness and efficiency criteria of the *system rationalities* that dominate the secondary collection phase. Domestic rationalities are influenced by cultural norms, by the domestic division of labour (also between men and women), to the domestic infrastructures, and to their daily and weekly rhythms.

As indicated in Figure 3.3 households are expected to bring their wastes to the communal transfer point, from where it is further transported to the disposal site. The bin, skip or transfer station can be regarded as points of connecting the primary and secondary phase actors and dynamics in handling domestic wastes. For that reason we will refer to this connection point as 'waste junction' (see Figure 3.3). This coordination and integration of households (downstream actor) and the secondary waste collectors (upstream actors), is of crucial importance for more effective and successful organization of the domestic solid waste system. When this coordination is established, it creates a stronger socio-technical network or chain for handling wastes. By doing so, it facilitates the development of cost-efficient and effective delivery services in ways as suggested by the MMA (Oosterveer and Spaargaren, 2010). In the primary phase of waste collection we see at work the kind of technologies and the forms of organization which are typical for the scale of operation of decentral systems, according to the MMA. They are characterized by their decentral/local characteristics and under the decentralized forms of management and control by households and local neighbourhood actors. The technologies and actors characteristic for the secondary phase in waste collection present another scale of operation which displays features of centralized system as defined by the MMA. These two types of technology and forms of organization can be said to meet at the transfer station. The centralized systems are mostly centrally managed by governments and little or no involvement of end-users. Decentralized systems on the other hand are small scale and flexible and allow high involvement of end-users (households) while preferably not relying on secondary, central scale level actors and technologies.

3.5 Conceptualizing the roles of households in domestic solid waste management chains

Figure 3.4 presents the conceptual framework that summarizes the discussion so far and that has been developed to guide and focus our research and to facilitate the formulation of research questions and their operationalization and implementation into empirical research.

The conceptual framework basically consists of four main parts. The upper part shows the position of households in the solid waste management chain. As can be seen in Figure 3.4 the second part shows the three basic roles of households and their functions in the SWM chain: (1) as solid waste generators, (2) as handlers of solid wastes in the primary phase of the chain, and (3) as recipients of solid waste management services. The household rationalities that influence waste management practices include among others: the division of labour (also between man and women), the prevailing cultural norms, the size of the household, the daily domestic routines

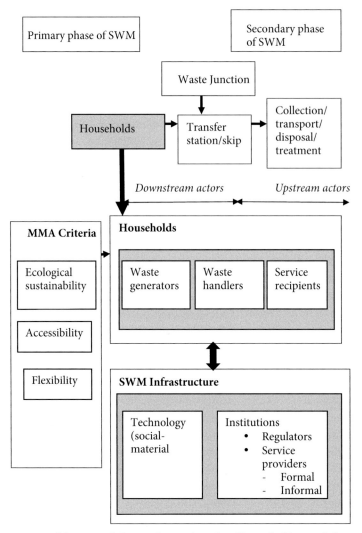

Figure 3.4. Conceptual framework for analysing the role of households in solid waste management practices.

within the household, and their basic perceptions towards waste management. The third part of the conceptual framework shows the SWM system as formed by the solid waste management infrastructures which include the (socio)technologies being applied and the relevant institutions for the organization and regulation of the chain. The socio-technological infrastructures include material as well as social elements, while the relevant institutions can be divided into the official service regulators and the actual service providers. The actual service providers in this context include formal and informal stakeholders. The fourth part of the framework – represented at the left hand side of Figure 3.4 – represents the criteria relevant for the assessment of (new) socio-technical infrastructures to be developed with the use of the MMA.

3.5.1 Households as key units of analysis

The developed conceptual framework makes households the central focus of attention for this study, while recognizing that SWM infrastructures have an important role to play in shaping and also improving SWM practices of households for example by directly influencing the choice of (collection and separation) technologies and by setting the price and other conditions for service delivery. Households, their practices and the infrastructures that are forming the 'context' for these domestic waste-practices should be analysed for their interdependencies.

As explained by Drangert *et al.* (2002) no environmental issue can be solved entirely by the city authority or by the state. And the involvement of the end-user is seen as a key factor to long-term sustainability since it provides a good way of getting to know and managing the expectations of users (Addo-Yobo and Njiru, 2006). The developed conceptual framework assumes that households as end-users of solid waste management services have a great role to play in solving the particular problems of solid waste. This gives the justification for placing household at such a strategic position within the framework.

The understanding of households' activities in SWM chain can provide important clues for SWM improvement. The activities associated with managing solid wastes from the point of their generation up to the point of final disposal and treatment can include generation, waste recovery (reduction, reuse, recycling, composting), waste storage, collection, transfer and transport, disposal, treatment. As pointed out by Adedipe *et al.* (2005) the flow of wastes from their place of origin to the site of disposal has important human dimensions to it, besides the application of waste technologies. The concept of a city or region functioning as an 'anthroposphere' indicates the importance of understanding households' activities in SWM chains.

Prerequisite for the effective management of domestic solid wastes is a good understanding of the households, that is, knowing what household think, do and the things that are influenced by what they think, and do. Addo-Yobo and Njiru (2006) stated that in order to get users of services to play their role effectively, there is a need to understand the beliefs and attitudes that underpin the decisions they make. For this reason, service providers and policy makers need to understand their citizen-consumers as end-users of services in order to be able to provide them with services in a satisfactory way and also to motivate citizen-consumers to play their required roles in ensuring a sustainable service delivery. Therefore it is important to pay attention to the households' perceptions and routines, because this could provide useful inputs for formulating waste management strategies and services that will underpin and serve the waste management practices of households.

Perceptions and social-cultural norms are vital to consider if the waste problem in informal settlements and densely populated areas in East African capital cities is to be effectively tackled. These cultural norms co-determine the ways in which individuals generate wastes and become involved in ways to handle and dispose of their wastes. Oosterveer and Spaargaren (2010) insisted that when it comes to developing sanitation infrastructures, particular cultural dynamics – for example norms about hygiene – are highly relevant to consider since they influence the (non) acceptance of new technologies and routines by households in a direct way. These cultural influences should also be considered when developing (new) solid waste management routines and systems at the household level.

Better understanding of the roles of households in SWM could help in developing strategies that will lead to the successful improvement of SWM at household level, thus enhancing the sustainability of solid waste services in general. Next to their cultural norms and perceptions, also basic characteristics of householders are relevant to consider. Household characteristics that influence SWM practices include the household size, the level of income, and the composition of the household. Finally, also knowledge about the kinds of domestic wastes generated and their composition will be an important element in improving waste management systems at the household level. Such knowledge about the waste itself helps in comparing the technologies to be chosen with the character of the wastes that are generated at particular locations and in specific (domestic) settings.

3.5.2 Solid waste management infrastructures

The SWM infrastructures as included in the conceptual model are specified for a technological dimension and a social dimension. The technology aspects here refer to the facilities such as storage containers, availability of skips, transfer stations, solid waste collection equipment. The assumption is that the ways in which households perform their roles in solid waste management chains depend very much on the technological and institutional set-ups provided by the infrastructures for domestic waste management. For example, transferring wastes to a transfer station or a skip is better possible when these skips are easily accessible by household (Barr *et al.*, 2001). When these facilities are accessible by households it can be more realistically expected that householders show their responsibilities for properly depositing their waste.

According to Anschütz (1996), the transfer station is one of the most visible aspects of SWM since it provides an interface between primary and secondary collection. For householders to carry their own waste to the transfer station, it needs to be located within (easy) walking distance. Only in this way indiscriminate dumping can be discouraged. Furthermore, Anschütz pointed out that when households often behave contrary to schedules and rules of effective solid waste management this can sometimes be shown to be caused by a lack of facilities. If a transfer point or dumping site is more than 100 metres away from their house, people tend to throw their waste much more often in streets, open spaces and rivers. The transfer station should be 'acceptable' to householders in terms of location, visual impact and method of disposing of waste. The availability of transfer stations is an important factor to be studied when trying to understand the deposition behaviours of households. Because householders are crucial for the ways in which waste-flows can be shown (not) to move between households and skips, the strong emphasis on households and the primary phase of SWM which is reflected in MMA seems to be justified.

As mentioned earlier, the institutional context includes the main institutions for dealing with solid waste management. They include the organizations of the stakeholders that are most relevant to household solid waste management, i.e. the decision makers, operators and regulators within the waste management system. The legal and institutional frameworks for their activities include the assignment of institutional responsibilities for wastes, existing guidelines, and policy and legal requirements pertaining to household waste management.

3.5.3 Modernized Mixtures Approach criteria: ecological sustainability, accessibility and flexibility

The concepts of ecological sustainability, accessibility and flexibility of solid waste management are reflected in the conceptual framework as indicated in Figure 3.4. In order to provide adequate solutions, Modernized Mixture approaches should be ecologically and institutionally sustainable, accessible (particularly for the poor), and institutionally and technically flexible, resilient and robust (Oosterveer and Spaargaren, 2010; Spaargaren *et al.*, 2005).

Accessibility reflects the extent to which specific groups within the urban population, such as women, poor or elderly are included or excluded from receiving sanitary infrastructures and services due to financial, physical, or cultural reasons.

The flexibility criterion points at the way a sanitation system might fit into more encompassing systems to be developed in the future, while also describing how the systems behave in times of instability of climatic, political, or economic nature. As pointed out by van Horen (2004) the solid waste sector is highly flexible also in terms of the ability to privatize some components and to encourage increased involvement of community and non-governmental sectors.

The sustainability requirement can be distinguished in institutional and ecological sustainability. Institutional sustainability concerns the extent to which a new system becomes embedded in existing socio-political and cultural systems at the local and national level, while improving their performance. Ecological sustainability refers to the achievements in waste prevention (reducing the need for final disposal of the waste) and reducing the demand for inputs, in particular water and energy.

Other authors have described ecological sustainability in solid waste management in almost similar perspectives to that of Spaargaren *et al.* (2005). Žičkienė *et al.* (2005) argued that in order to achieve ecological sustainability in waste management, the following goals need to be realized: (1) production of waste should be minimized through new organization of production processes; (2) reuse and recycling should be maximized; (3) remaining waste should be disposed of in a controlled way in order not to exceed the absorption capacity of local sinks. Reuse and recycling make essential contributions to ecological sustainability in several ways: (a) the demand for natural resources is reduced, (b) emissions to environment are decreased (less energy is used for reprocessing secondary materials than for extraction of virgin materials), and (c) the amount of the solid waste is reduced and smaller amounts of waste remain for disposal. Baud (2004) mentions three goals with respect to ecological sustainability. First, the aim is to minimize the amount of waste generated, second to minimize reuse and recycling and thirdly to dispose of remaining waste in a controlled fashion in order to not exceed the capacities of the local sinks.

While in the literature several definitions of (ecological) sustainability, accessibility and flexibility are available, we think the criteria as derived from the MMA are relevant and useful for analysing and (re)designing and also improving the practices, infrastructures and institutions involved in the management of domestic solid wastes in East African capital cities. The criteria serve as normative guidelines when assessing technologies, practices and also decision making procedures concerning domestic solid wastes.

3.6 Research questions

The roles and responsibilities of household in domestic solid waste management have been theoretically explored, using the body of literature on solid waste management as main point of reference. The roles of household were specified into three main dimensions. First as waste generators, second as key actors for waste handling in the primary phase of the waste-chain and third as recipients of municipal solid waste services. These roles were explored in some detail, resulting in the conceptual framework guiding our empirical research. Based on this conceptual framework, the following specific (operational) research questions and sub-questions can be formulated:

1. What are the characteristics of household wastes?
    a. What is the per capita generation and composition of household solid waste?
    b. What are the factors influencing the waste characteristics at the household level?
2. What are the key domestic practices and routines for handling solid wastes in the primary phase, at the household level?
    a. What are the current waste management practices applied by households?
    b. How do the domestic solid waste flows 'travel' from the households to the transfer station?
    c. What are the different methods that households use in managing waste?
    d. What are the roles of different household members in the practice of waste management?
3. How do households receive solid waste management services and what are their basic perceptions and evaluations of the ways of being served?
    a. Who are the key formal and informal stakeholders in the waste chain for domestic solid waste management?
    b. What kind of formal and informal relationships exist between householders and other stakeholders in the SWM chain?
    c. What forms of institutional support do exist for the relationship between householders and other stakeholders in the SWM chain?
    d. How do householders perceive and evaluate the actual waste services as resulting from the formal and informal relationships they uphold with other stakeholders?

These questions will be used to organize the empirical research. They are also used to structure the chapters on the empirical results. Before we start reporting on the empirical results however, we will first discuss the research methodology in some detail in the next chapter.

Chapter 4.
Research methodology

4.1 Data collection methods

So as to answer the research questions a combination of both primary and secondary data collection methods was employed. In order to explore household waste management in depth, a comprehensive empirical study was undertaken covering low-income households in Kinondoni municipality in Dar es Salaam city, Tanzania.

As mentioned in the introduction of this thesis, the study was partly carried out in Nairobi and Kampala in order to understand solid waste management from household perspectives from other capital cities of East Africa. This comparative research methodology first and for all was used to become aware of some particular characteristics of our Dar es Salaam situation. The cities used for comparison are the largest and fastest growing cities in East Africa where 50-70% of population live in informal settlements with inadequate solid waste management services. Dar es Salaam[13] is the principal commercial city of Tanzania and the de-facto seat of most government institutions and it is estimated that 60-70%, of its population live in informal settlements. According to Kyessi (2005), 70% of its population live in informal settlements. The city of Dar es Salaam consists of 3 municipalities i.e. Kinondoni, Temeke and Ilala. Kinondoni was purposively selected because it is the largest municipal council, and fastest growing municipality with a population of 1,088,867 and growth rate of 5.4% (NBS, 2002).

Nairobi is the capital city of Kenya with population of 3.5 million, where 60% of its population live in informal settlements. Whereas, Kampala is the capital city of Uganda with over 1 million people making it Uganda's largest city, where about 57% live in informal settlements. The city of Kampala is divided into five administrative Divisions: Central, Kawempe, Makindye, Nakawa and Rubaga. Kawempe division was purposively selected because Kawempe is the most densely populated suburb of Kampala compared to other divisions and is greatly affected by poor solid waste management. The comprehensive research in Dar es Salaam involved both qualitative and quantitative methods, while in Nairobi and Kampala the research was limited to qualitative methods only.

Qualitative and quantitative methods are complementary, and their combined application increases both reliability and validity (Niehof, 1999). In this study a combination of qualitative and quantitative methods of data collection were employed to gain a deeper understanding of the household solid waste management in its current state, in order to devise viable solutions towards improving. The information was collected from the level of central government which was represented by officials from environmental councils, from local government officials and the household level. Table 4.1 presents the key informants from each city selected to provide qualitative information. A total of 18 individuals were selected as key informants and they included: National Management Environmental Councils officials, local government officials and legal solid waste

---

[13] Tanzania officially designated its capital as Dodoma, but only the legislature meets there leaving Dar es Salaam as the de facto capital city

Table 4.1. Key informants interviewed from each city (number of informants between brackets).

Key informants	Dar es Salaam	Nairobi	Kampala	Total
Central government official – Environmental Councils officials	NEMC (1)	NEMA (1)	EMA (1)	3
Local government officials	Dar es Salaam City Council health officer (1); Kinondoni Municipal health officer (1)	Nairobi City Council officials (1)	Kampala City Council – engineer dealing SWM (1)	4
Solid waste collection providers	Waste Contractors (CBOs, private companies) (3)	CBOs (4)	Private waste contractors, NGOs (4)	11
Total	6	6	6	18

management contractors (CBOs and private companies). The quantitative information will be described in section 4.2.6.

The rationale for conducting interviews with these people was to obtain expert information on solid waste management, given their knowledge, and experience. The different key informants were purposively selected based on the following criteria: National Management Environmental Councils officials, are responsible for policy formulation; local government officials, have overall responsibility in solid waste management; and waste contractors are solid waste management service providers.

In Dar es Salaam key informants were the Dar es Salaam City Council health officer, the Kinondoni Municipal Council health officer and the Director of Tanzania National Environment management Council (NEMC). Dar es Salaam City Council (DCC) has overall responsibility of the waste management of the city. An official from NEMC was selected because NEMC is responsible for policy formulation and advising the government on environmental issues in general. The selected waste contractors included: KWODET from Hananasifu sub-ward, CLN from Makangira sub-ward and TTM from Kilimahewa sub-ward. These waste contractors were selected because they provide solid waste management services in areas selected for this research. In total 6 key informants were selected.

In Nairobi key informants included the acting director from the Environmental department in Nairobi City Council (NCC), an official from the solid waste management division, an official from NEMA and selected solid waste contractors. An official from NEMA was selected because NEMA is the principal instrument of government in the implementation of all policies relating to the environment, while an official from Environmental department was selected because the Department of environment has an overall responsibility in solid waste management.

Solid waste contractors selected for interviews in Nairobi were purposively sampled from SACCO. SACCO is a cooperative which draws its membership from members of various CBOs of Nairobi, its goal is to engage in economically viable environmental friendly and profitable community managed plastics waste streams in the settlements of Nairobi city. During the period of this research SACCO had registered 44 members. Identification and sampling was done with the help of a cooperative chairman from a list of cooperative members using a list of registered CBOs as a sampling frame. Consequently, 3 active CBO members were selected. The selected CBOs provide service in slum areas Mbotela, Dagoretti and Muthurwa. The selected waste contractors included: Kamaliza youth group, Riruta Environmental group and Muthurwa Eco Club. In total 5 key informants were selected for Nairobi interviews.

In Kampala key informants included Kampala City Council Officials, NEMA and an Engineer responsible for SWM in Kawempe Division. The private solid waste contractors were purposively sampled from a list of waste collection and transportation firms operating in Kampala which was provided by KCC Engineer. Out of 16 firms operating the whole of Kampala, Kawempe division is served by two firms namely NOREMA Services Ltd and Hilltop Enterprises Ltd. Two NGOs, Plastic Recycling Industry and Envirocare Initiative, both active in recycling activities based in Kampala were also interviewed. The two NGOs were interviewed on the advice of Kawempe Solid waste management Engineer because they closely work with waste contractors who provide solid waste collection services to households.

4.2 Methods of primary data collection

Qualitative methods used included semi-structured interviews with key informants, direct observation and focus group discussions. The quantitative method used was the household survey questionnaire. The following sub-section explains each method from the point of view of its applicability in this research.

4.2.1 Reconnaissance visit

The pre-fieldwork phase of the research involves the reconnaissance visit. It involved a preliminary survey which was conducted by a researcher in September and October 2007 in Dar es Salaam city. The reconnaissance visit helped to identify the research areas and gather relevant preliminary information on the areas. Objectives of the visit were to establish contacts with central and local government officials dealing with solid waste management and look at the possibilities of cooperation during the research period. Through this visit information obtained mainly from several secondary sources of available publications, books, reports, working papers, policy documents, etc. were consulted. Discussions with solid waste management officials and informal discussion with other experts in SWM assisted in establishing criteria for sampling and in the selection of areas where the survey was to be conducted. Subsequently primary data were collected. Table 4.2 provides the information about the data collection methods, the main purpose and the type of information obtained from each method.

4.2.2 Interviews with key informants

Semi-structured interviews were an important method used in this research, to allow flexibility towards specific respondents and two-way discussion with respondents while maintaining focus

Table 4.2. Summary of data collection methods and corresponding objectives.

Data collection techniques	Respondents	Study object
1. Reviewing documents and reports available on waste management (quantitative data and qualitative data) by a researcher	Researcher	Review of policy documents to obtain general overview of SWM
2. Observing (qualitative data) by a researcher	Households	To get the actual condition of the research objects
3. Face to face interviews (qualitative data) by a researcher	Central government officials	Information on policy and legal framework, and verify secondary information and data documentation
	Local government officials	Understand the existing SWM scenario, discuss SWM related problems, explore their views regarding household waste management practices, verify information and data reviewed during the desk top study
4. Face to face interviews (qualitative data) by a researcher	Waste contractors	Acquire information on their roles and daily activities in SWM. Explore their views and suggestions on the improvement of HSWM
5. Focus group discussions (qualitative data) by researcher, recorders, and moderators	Households	Obtain views and opinions regarding our main study objective
6. Questionnaire survey (quantitative data) by research team / Waste characterization (quantitative data) by research team	Households	Information on household SWM practices, waste flow to transfer station, households perception towards waste management provisioning, determine per capital daily waste generation and waste composition

on the major issues to be discussed. The guide for conducting the interviews was developed by the principle researcher to at least allow the researcher to obtain data within the designed scope of the project. The interview guide was used by the researcher to ensure all the relevant areas were covered. The interviews started with a basic introduction of research objectives before issues about solid waste management were discussed. The topic list on issues to discuss was prepared in advance of each visit as recommended by Stewart and Shamdasani (1990) and Stewart *et al.* (2007). All interviews were conducted in person (face to face) by a researcher, taking notes and probing questions based on what respondents said. Interviews were transcribed immediately after the interview for further processing and issues requiring clarification were verified with respondents though re-visits. Interviews were conducted in both English and Swahili commonly spoken by the majority of the respondents. Individual interviews lasted for 45 minutes to 1 hour which is consistent with Yin (2003) who indicated that interviews may take 1 hour for the maximum.

The first interview was conducted with Dar es Salaam City health officer. The information obtained was generally on an overview of waste management in Dar es Salaam. From this interview the stakeholders involved in household waste management were also identified. The second interview was conducted with Kinondoni Municipal Health Officer in order to identify the current situation of solid waste management, and to learn the institutional support to household waste management, as well as obtaining views on how to improve households' waste management. All local issues affecting solid waste management was taken into consideration including: by-laws in place, policies and regulations, roles of other stakeholders (public sector roles, private sector), and institutional arrangement. The interview with Kinondoni health officer provided information for research questions 1, 2 and 3 (Chapter 3.6) of this study. The third interview was carried out with the director of the National Environment Management Council (NEMC). This interview was carried out because NEMC is responsible for policy formulation and advising the government on environmental issues in general.

Interviews were also conducted with solid waste management service providers (CBOs and private companies) in areas selected for research. The main objective of these interviews was to acquire information on their roles regarding SWM and views on household waste management. The interviewed waste contractors included: KWODET from Hananasifu sub-ward, CLN from Makangira sub-ward and TTM from Kilimahewa sub-ward.

Nairobi interviews were conducted with the acting director from the Environmental department in Nairobi City Council (NCC), an official from the solid waste management division, an official from NEMA and selected solid waste contractors. The aim of interviews with the acting director from the Environmental department and NEMA was to obtain an overview of solid waste management in the cities, and the service provision to households in slum areas.

The interview with an official from Environmental department was carried out because the Department of environment has an overall responsibility in solid waste management. The aim of the interviews with waste contractors in Nairobi was to obtain information regarding waste management services to households they serves. The information included the roles of households, mode of service provision and the problems encountered.

In Kampala interviews were carried out with Kampala City Council Officials, NEMA and an Engineer responsible for SWM in Kawempe Division. The aim of the interviews with waste contractors in Kampala was to obtain information regarding households' waste management services to households they serves. The information included the roles of households, mode of service provision and the problems encountered.

4.2.3 Focus group discussion

The focus group approach was selected as it provides a fast and efficient manner to get a preliminary feel of the resident's views on local problems and priorities in a sphere of trust and meaningful interaction (Morgan, 1993). The summary of the elements of focus group sessions are presented in Table 4.3

Table 4.3. The elements of focus group discussion for this research.

Place	Date	Venue	No[1]	Time	Communication	Format
Dar es Salaam	8 January 2009	Zanaki Secondary school	12	10:00-12:10	Swahili	open session
Nairobi	15 January 2009	SACCO Meeting hall	14	10:30-12:30	Swahili/English	open session
Kampala	19 June 2009	Akamwesi hostel	12	11:00-13:15	Swahili/English/Baganda	open session

[1] No = number of attendants to the group discussion.

The participants were selected so as to ensure there is a degree of homogeneity in the group. For the Dar es Salaam focus group, a sample was randomly selected from household questionnaire sample. The Nairobi and Kampala the participants were drawn from areas served by CBOs/private companies involved in interviews. The local leaders were used as intermediary persons to select participants for the focus group discussion. Participants were identified by purposive selection, to include those who are open and could provide information in solid waste management and have experience in their locality. Gender balance was another selection criterion of the participants. In terms of age, we included participants ranging between 20 to 65 years old. To ensure attendance, invitation letters were sent to each participant 2 weeks before the date of the focus group. The letters stated the objective of the meeting and the intended topics of discussion. Discussion lasted for about 90 minutes to 2 hours.

According to Lewis (2000) numbers of participants can range between about 6 to 12 participants. In this study, in total 3 focus group discussions were conducted involving 10 to14 per group, this number allowed each person in the group to have a turn to speak, and so that sub-groups were not formed, it become easier to moderate and gave the participants adequate time to express their views and opinions. The main steps involved in focus group discussions for this research were based on the frameworks provided by Krishnaswami and Ranganatham (2007) and Kreuger (1988, 1994). The method takes advantage of the interaction between small groups of people. Participants respond to and build on what others in the group have said. For this study the focus group discussions involved householders who represent households in the studied areas.

The objective of the focus group session in this study was to create a better understanding on current household SWM from the perspectives of the households. Their perceptions and opinions regarding existing solid waste management were examined to obtain views on services provisioning by formal and informal waste providers, and to give opinions on their own roles in solid waste management in order to identify the future and potential options in solid waste management. The focus group discussions were carried out using a discussion guide which contains question which led the discussion (see Appendix 6). The guide was developed in Dar es Salaam by the researcher assisted by the research supervisor and a chosen moderator. It was developed in English and translated into Swahili. Moderators in Dar es Salaam and Nairobi communicated effectively in Swahili that is commonly spoken, while in Kampala a moderator communicated in English, and was translated to a local language with the aid of an employed translator. The Dar es Salaam focus group discussion was moderated by the researcher from Dar es Salaam Institute of Technology who has had experience in SWM and in researching solid waste, and also has an experienced qualitative researcher who had run focus groups for other studies. The PROVIDE project research fellows were the moderators in the Nairobi and Kampala focus groups. The moderators were chosen as they were very familiar with the topic of discussion and thus able to put all comments into perspective and follow up critical areas, as was required in this research. The role of the moderators was to facilitate the group interaction, to encourage the group, and ensuring that no individual participant dominates the discussion and leads discussion through the range of topics of interest to the end.

4.2.4 Direct observation

Direct observations were used to document the actual conditions on household waste management to corroborate the responses from interviews and questionnaire surveys. Direct observation, with the aid of checklist, involved watching what is happening, and recording events on the spot.

The observation was loosely designed to include things to look for which include type of waste generated by households, storage of waste, collection methods, who is providing the service, who is responsible for waste handling at household level, what they do with waste, characteristics of households. Such guided observation helped to focus on solid waste management practices at household level. Observation was undertaken by the researcher. Two research assistants participated in the beginning, but they failed to record the events and were dropped from the exercise. This data collection technique was employed to collect information prior to the other methods, and throughout the period of research. The observed households were selected from the large sample of households which participated in questionnaire survey. As direct observation is time consuming it is most often used in small-scale studies. Following this fact, ten randomly selected households were visited and observed thoroughly each for two days in a week.

Sub-ward leaders in the respective sub-wards were involved in accessing the selected households. However, the researcher was observing each time she visited the research areas and other various households, during interviews, questionnaire survey, during walks, family and friends visits. Sometimes the researcher stayed in strategic position such as sitting in a corner and observes activities of households in regard to solid waste management for hours in order to get a clear picture. The researcher also observed the solid waste flow to transfer stations. Field notes were made while observing, and soon after the observation were processed.

4.2.5 Secondary materials

Various policy documents were used as a source of secondary data. After obtaining primary data through interviews, various documents were reviewed to supplement the information from interviews. Documents reviewed included the local government (Urban Authorities) Act 1982, Kinondoni municipal waste management by-laws, waste collection and disposal by-laws (2000 and 2001) of Kinondoni Municipal Council, management of solid wastes in Dar es Salaam, National Environmental Management Council Tanzania (1996 – unpublished report), the United Republic of Tanzania, Ministry of Health Waste Management Guidelines 2003, Guidelines and Performance Standards for Environmental Health Personnel 1998, National Environment Policy Document of

1997 and National Health Environmental Health and Sanitation Policy Guidelines (Ministry of Health, 2004). More information on secondary material was obtained through intensive reviews of literature and various publications related to solid waste management.

4.2.6 Quantitative data collection

Quantitative data were collected through a structured household survey questionnaire undertaken between March 2008 to July 2008. The objectives of carrying out household survey questionnaire was to collect information about households knowledge on solid waste management practices at household level, waste flow from households to transfer station, to obtain households' perceptions towards solid waste management provisioning. This study was carried out in 6 sub-wards under the jurisdiction of the Kinondoni Municipal Council in Dar es Salaam city. Table 4.4 presents names and the main characteristics of the selected sub-wards.

Sampling procedure

A combination of purposive sampling, clustering and random sampling methods was used in selecting the study areas in Kinondoni municipality. Administratively, Kinondoni municipality comprises wards, sub-wards and households as the lowest level of governance. Kinondoni municipality is divided into 27 wards, which in turn are divided into villages for rural areas and sub-wards commonly known as *Mitaa* in the urban areas. There are 113 sub-wards. The greatest concentration of poorly serviced unplanned settlements is located in Kinondoni.

Accordingly, the three sub-wards of Hananasifu, Kilimahewa and Makangira are categorized as low-income sub-wards, while Ubungo-Kibangu, Mwongozo and Makoka are categorized as middle-income sub-wards. However, the income category was not the basis of selection. These settlements were selected for the study because they represent areas that are predominantly informal in nature and have relatively high residential density. The settlements are characterized by a lack of basic infrastructure services and overcrowding. Documentation produced under the Sustainable Dar es Salaam Program (SDP) indicates that these settlements are unplanned and generally unserviced (World Bank, 2002b), and in total there are 35 unplanned settlements.

Table 4.4. Main characteristics of the study sub-wards (adapted from Kinondoni municipal Council, 2008).

Sub-ward	Population density (persons per hectare)	Location (distance to CBD in km)	Income category
Hananasifu	>2,000	4 north	low-income
Kilimahewa	1000-2,000	6 north west	low-income
Makangira	>2,000	3.5 north east	low-income
Ubungo-Kibangu	>2,000	5 north west	middle-income
Makoka	>2,000	4.5 north west	middle-income
Mwongozo	>2,000	4 north west	middle-income

Selection of households

From each selected sub-ward clusters were formed as presented in Table 4.5. The sampling frame was provided by the respective sub-ward leaders of each selected sub-ward. An effective sampling interval of 100 was applied to obtain clusters of households in sub-wards. A systematic random sampling procedure was employed to generate a sample of 360 households in total. To get representative results in the survey it was decided to select 5 households randomly from the clusters picked for the first three sub-wards, and 3 households from the clusters picked for the last three sub-wards as indicated in Table 4.5.

Table 4.5. Number of households, clusters and sampled households in selected research areas (based on information from Kinondoni Municipal Council, 2008).

Sub-wards	Number of households	Clusters sampled	Number of households studied
Hananasifu	1,420	14	70
Kilimahewa	1,349	13	67
Makangira	850	8	43
Ubungo-Kibangu	2,271	21	42
Makoka	3,311	33	66
Mwongozo	4,000	36	72

The sample

The characteristics of the respondents are as shown in Table 4.6. Based on the sample of 360, 81 (22.5%) respondents were male, while 279 respondents (77.5%) were female. In terms of occupancy status, the majority (71%) were house-owners, whereas 29% were tenants. The respondents were aged between 21-77 years, with the majority (81.9%) between 29 to 55 years. 10.6% represented ages between 21 to 28 years, and the remaining 7.5% between 56 and 77 years. The mean age of respondents was 40 years. The source of income for the majority (185 respondents; 51.5%) was informal employment, 153 respondents (42.5%) had formal employment, while 22 respondents (6%) could not specify their source of income. About 56% had primary school level education, about 36.5% had secondary school level education, while 7.5% had tertiary education.

With regard to income levels (Table 4.6), 23 respondents earned between TZS 0 to 50,000 per month, 117 respondents earned between TZS 50,001-100,000 per month, 60 respondents earned between TZS 100,001-150,000 per month, while the majority (150 respondents) earned more than TZS 150,000-200,000 per month. The remaining 10 respondents earned more than TZS 200,000.

Table 4.6. Respondents' characteristics.

Gender	male	female			
	81 (22.5%)	279 (77.5%)			
Occupancy status	house-owners	tenants			
	256 (71%)	104 (29%)			
Age	21-28	29-55	56-77		
	38 (10.6%)	295 (81.9%)	27 (7.5%)		
Source of income	informal employment	formal employment	not specific		
	185 (51.5%)	153 (42.5%)	22 (6%)		
Education	primary	secondary	tertiary		
	202 (56%)	131 (36.5%)	27 (7.5%)		
Income (TZS)	0-50,000	50,001-100,000	100,001-150,000	150,000-200,000	>200,000
	23 (6.4%)	117 (32.5%)	60 (16.75%)	150 (41.55%)	10 (2.8%)

Household survey

The developed questionnaires were pre-tested and revised before being administered by the researcher and two research assistants. The pre-test exercise was implemented in Hananasifu sub-ward during period of one month (January 2008), after which all appropriate adjustments were made and final questionnaires produced. Pre-testing was done in consistence with Krishnaswami and Ranganatham (2007) to find out whether households were comfortable with the questionnaire, answers provided the information required, if all words were understood, and to determine the time it took to administer.

The questionnaire in English version was translated into Swahili. The pre-testing of the Swahili version was conducted systematically using respondents (households) drawn from a large sample of survey using the same method of administration. Revisions were made based on their input and changes were made accordingly. A final English version was produced based on the final Swahili version.

A closed format approach questionnaire was adopted in this research combined with some open format questions. The open-ended questions were included for the purpose of getting the deep answers from the respondents. Closed format questionnaires are easy to ask and quick to answer and it is easy to calculate percentages and other statistical data over the whole group or over any subgroup of participants

The research team administered pre-tested and revised questionnaires written in English and translated in Swahili. Households were informed about the purpose of the survey before the actual exercise. In order to ensure the highest levels of compliance and cooperation from households, the research team identified the key contact persons from the area before administering of the questionnaire. Key contact person were, sub-ward leaders and a CBO/Private company providing services in a respective research area.

The final questionnaire contained 27 questions grouped into 5 main parts: The following attributes were included in the final questionnaire:

- Part 1 contained question 1-4 on demographic information which is concerned location and size of household, characteristics and status of respondents in terms of sex, marital status and age. Socio-economic status variables include education, source of income, monthly income occupancy status. Another variable which is included in this part is business activity. These variables may affect solid waste management practices, amount of waste generated, attitudes and perception of the household towards solid waste management. The demographical information was also used to provide background for the data analysis.
- Part II contained questions 5-9 concerning the knowledge on solid waste management. This part was included in order to learn what households already know about solid waste management.
- Part III contained questions 10-18 concerning with the households' solid waste management practices. Responses to these questions create a better understanding of waste storage, waste separation, the daily routine of households and the responsibility of household members in solid waste management.
- Part IV contained questions 19-22 concerning with solid waste collection and disposal. Responses to these questions are used to gain insight on waste flow to transfer stations by looking at who is providing the collection service, the schedule and frequency of collection, how households take part in primary phase of waste collection.
- Part V contained question 23-27 concerning with the households' views on the services provided to them. Responses to these question help us to understand how household would like to participate in SWM.

Data analysis

The analysis of data collected from questionnaire survey was entered into the computer using Microsoft Access. Data entry was manually cross-checked to ensure accuracy. Descriptive statistics such as means, frequencies, and standard deviations were computed by the use of the Statistical Package for Social Sciences (SPSS version 16). A bivariate analysis using Pearson's coefficient (r) was used to find the correlation between household size and per capita waste generation, and also between income of the households and per capita waste generation.

Data collected from interviews, and focus group discussion were processed and edited as follows: these data were recorded by taking notes and transcribed before being analysed using content analysis. Content analysis is a method for analysing textual data expressing key ideas, phrases and meanings in answers given to interview questions (Weber, 1990).

To handle data carefully, discussions notes were recorded by the researcher and a second person during interviews and focus group discussions. The notes from a researcher and a recorder who participated in each discussion were combined for a comparison, and any discrepancies in their notes resolved between them. In addition comparing notes with the second recorder enhanced the reliability. Personal judgments, comments from experts, and results from interviews were also used as a basis for the analysis and interpretation of the information.

4.3 Waste characterization study

Waste characterization was carried out by the research team to determine the per capita generation per day and the physical composition of the household waste. Table 4.7 present the summary of the materials and procedure adopted in waste characterization study.

The determination of generated waste was performed in the households where also questionnaires were administered. Measuring of waste components took place on the spot. Determination of waste composition was done by way of physical separation and observation

Table 4.7. Summary of materials and procedure adapted in waste characterization.

Materials used for measurement of waste in order to determine the generation rates and waste composition

- A number of plastic bags with volume of 12 litres for collecting and weighing the waste
- Rope to tie the neck of the bags to prevent the waste from falling out
- Portable weighing balances of capacity 50 kg and 100 kg for weighing the waste
- Sufficient recording sheets
- Shovels, forks, gloves, facemasks for personal protection
- 4-6 buckets for waste sorting
- Plastic sheet for waste analysis (4 m^2)

Procedures adapted for waste characterization
- Ten empty plastic bags were weighed to determine the average weight of one plastic bag
- Plastic bags (2 to 3) were provided by research team to all selected households, the purpose of the study was explained and households were asked to store their waste generated for the particular days of study in the bags provided
- The research team returned at every alternate day to weigh and sort each sample bag assigned to households and record the weight into the data sheet provided. Waste was sorted into pre-determined seven fractions namely food waste, glass, plastics, paper, metal/tins, aluminium, and others (inert and ashes)
- The contents of each bag of waste were weighed first and its weight recorded. Calculation on the per capita waste generation was obtained by dividing the total amount of waste generated per household by number of household members (household size). Then the contents were emptied and spread onto a plastic sheet and separated
- Sorting was carried out on the spot
- The separated waste was transferred into different plastic bags and measures the weight of each category of waste fraction separately and records its weight in the data sheet. Calculation on the household waste composition was performed
- All the waste was filled into the bags for a proper disposal and equipment were washed for next weighing
- The above steps were repeated for each selected household

of the collected waste. Each selected household was provided with a plastic bag to keep all their waste generated for the particular day of study. One for each day of the study, and one extra bag in case this was necessary. In order to obtain a realistic estimate, measurement of the amount of waste produced by a particular household was performed for three different days, in a week and then the average value was recorded. The amount of waste generated per capita per day for each household in the selected study area was determined. The term per capita refers to the waste generation per person.

An earlier methodological approach was designed to request households to sort waste they produce as per direction, but this approach did not work out. First, 10 selected households were each given 8 plastic bags and instructed to separate their waste into seven fractions, all of them failed to pre-sort the waste as required, hence the procedure was modified. The new approach required households to keep their waste into one storage bag for the research team to sort and weigh. They claimed it was a very difficult exercise for them and time consuming. However we speculated that respondents viewed it as dirty and unpleasant work, though they did not disclose this to us. The research team weighed the stored waste from each household and sorted it into seven pre-determined fractions, namely: food waste, glass, plastics, paper, metal/tins, aluminium, and residues (inert, ashes, and sweepings). During characterization the waste composite was physically observed to identify the components. Each waste fraction was weighed separately and its weight recorded on a sheet prepared by the researcher. Weighing and sorting of household wastes at the source makes the identification of waste materials easy and eliminates any uncertainty as to their origins (Oyelola and Babatunde, 2008). Table 4.8 describes the definition of different types of waste generated by households adopted in this study.

Table 4.8. Definition of waste types generated by households (Adapted from Tchobanoglous et al., 1993).

Waste type	Composition
Plastic	containers for oil, packaging, tubes of toothpaste, shampoos, bags, rubber
Metal	fruit and vegetable containers, components of machines, vehicles
Glass	broken glasses(bottles, window)
Paper	newspapers and magazines, packaging, books
Kitchen waste	food left-over, vegetable and fruit peels
Aluminium	cans of soft drinks and beer, food cans
Residues	ash, dust, silt, sweepings, pieces of wood, charcoal

4.4. Training of research assistants

A training session with the interviewers helped to ensure that research assistants understand all instructions and all questions on the survey. They were guided on how to code specific questions and how to respond to specific issues that are likely to arise during interviews (some of which may have been experienced in pre-tests).

Training of the research assistants was conducted at Dar es Salaam Institute of Technology premises. The research team involved 5 interviewers. Training lasted for one week and covered the following: (1) the research objectives; (2) the research instruments; (3) data collection techniques; and (4) data editing consistency checks in the questionnaire. A researcher ensured that a common understanding of the issues was developed amongst research team members. Common training and instruction was meant to reduce interviewer effects in the surveys.

4.5 Research reliability and validity

The reliability and validity address issues about the quality of the data and appropriateness of the methods used in carrying a research project. A number of measures were taken to ensure reliability and validity of this study.

First, the use of multiple methods (triangulation) in order to corroborate data sources increased the reliability of the research. The idea behind triangulation is that more agreement of different data sources on a particular issue, the more reliable the interpretation of the data. This ensures that the issue is not explored through one lens, but rather a variety of lenses which allows for multiple facets of the phenomenon to be revealed and understood (Baxter and Jack, 2008). In this research direct observation, existing records and in-depth interviews were used to triangulate the findings from the questionnaires and interviews. According to Yin (2003), interview data are corroborated with information from other sources to increase the validity of the study. The information obtained from government, city, and municipal officials was supplemented by information from written documents, and the data from the solid waste contractors were triangulated by the direct observations by the researcher.

The researcher asked others for comments; the experience, views and opinions were shared with other experienced researchers. Trained research assistants who could control the quality of the results and communicate in local language in the selected research areas were used. To ensure good responses from the respondents, the interviews were carried out in quite places. The questions were kept simple and very clear, the interviews and focus group discussions were short to avoid distractions and questions were asked in a logical order using a guide. In addition, comparing notes with the second recorder enhanced the reliability of the focus group information. To increase validity a researcher was verifying findings by taking the transcripts back to some of the interviewed participants to get more clarification on the information collected.

In order to ensure the highest levels of compliance and cooperation from households, the research team identified the key contact persons from the area before administering of the questionnaire. Key contact person were ward leaders, sub-ward leader and a CBO/private company providing services in a respective research area.

To minimize error possibilities, the study engaged the use of monitoring through visitation by research team leader (researcher). The author was present throughout the questionnaire survey in order to clarify uncertainties expressed by the respondents and pose additional more specific questions as required.

Chapter 5.
Household solid waste characteristics

5.1 Introduction

Households are the main focus of attention in the context of this study. The Swahili word for household is '*kaya*' which is used in the studied areas as well. As indicated in the conceptual framework presented in Chapter 3 of this thesis, households were presented in solid waste management chain as solid waste generators, as handlers of solid waste in the primary phase, and as recipients of solid waste management services. In order to improve households' solid waste management in the selected study areas, it is first and foremost important to understand the three roles of households in SWM chain. Building upon the findings achieved, insights can be gained with respect to the future improvement of solid waste management both in the social and technical dimensions of the process. The current chapter presents the study of households as waste generators, while Chapter 6 and 7 present the study of households as waste handlers and service recipients respectively.

It will be helpful to describe the meaning of household waste as given by different authors and as applied in this study. JICA (1997) defined household waste as waste arising as a result of domestic activities including food preparation, sweeping, cleaning, fuel burning, gardening, garbage (e.g. old clothes, newspapers, obsolete appliances, etc.). Kiely (1997) stated that household waste does not include waste products from agriculture, industry, mining and commerce. According to Hedén (2001) domestic household wastes can be solid (paper, plastics, food wastes, ash, glass, metal, etc.), liquid (old medicines, spent oils, paints, etc.) and also hazardous wastes (batteries, chemicals compounds, spent oil, etc.). Tchobanoglous *et al.* (1993) report that hazardous waste poses a substantial present or potential hazard to humans or other living organisms. However, hazardous waste is not dealt with in this study.

For the purpose of this study, household solid waste (HSW) is defined as the wastes generated by the domestic activities and non-domestic activities discarded by householders in the household. For example, in Dar es Salaam many householders were observed using their household premises as a site of preparing food which is later sold in another location. Another example is papers and plastic materials which are used as packing materials to bring commodities home, and discarded after being used together with other domestic wastes such as kitchen waste and sweepings.

As household waste contributes for a large part to the total amount of solid waste generated in different East African capital cities, there is a need for household waste to receive the appropriate management services to improve the health of both the public and the environment. The study carried out by Kasseva and Mbuligwe (2005) in Dar es Salaam showed that the households alone generate about 56% of the total waste generated. In Nairobi it is estimated that household waste contributes to about 83% of total municipal solid waste (NCC, 2008), while in Kampala the estimated contribution of domestic waste to municipal solid waste is 52% (Okot-Okumu, 2006). These results set out the necessity for household waste to receive the appropriate collection and disposal services for the better health of both the public and the environment.

Appropriate solid waste management provides opportunities for reducing environmental degradation and other associated problems (Mwai *et al.*, 2008). The first stage in dealing with solid waste is to understand the nature of waste being generated. Based on this fact, this study starts with exploring the constituents of waste generated by households. The types of waste being generated, its composition and distribution are the basic data needed for the planning (Ahmed and Ali, 2006). Household wastes can be extremely variable in their composition, depending to a large extent on the lifestyles of the generators. For example, it can be expected that in those countries where almost everything bought is associated with wrapping materials, the packaging waste very often comprises a significant part of household wastes. There will also be foodstuffs adhering to it or unusable material derived from foods preparation, such as vegetable peelings, meat scraps and bones, which make it unattractive for recycling.

The objective of this chapter is to create a better understanding of the daily per capita generation of household waste, the percentages of various components of household waste in the studied areas, and the factors affecting the two components. A complementary purpose of the study is to add to the limited amount of research on household waste in East Africa. This part of the study is guided by the following research questions:

1. What are the characteristics (per capita generation and composition) of household solid waste?
2. What are the factors influencing waste characteristics at household level?

This chapter is organized as follows: section 5.2 contains the conceptual framework. Section 5.3 contains the methodology of the study. Section 5.4 presents the results of the waste characterization and section 5.5 the factors which affect the solid waste characteristics. Finally section 5.6 presents the conclusion and discussion.

5.2 Conceptual framework

In Chapter 3 the conceptual framework guiding this research was introduced. Based on the developed conceptual framework, households are placed at the core of domestic solid waste management chains, because households are the locus where the main end-users or recipients of solid waste management services are located. The end-users are at the same time the primary producers of solid waste and the key actors in the primary phase of the waste management chain. Being at the origin or the down-stream end of the SWM chain, they determine daily waste generation and the types of waste generated. This chapter discusses the waste generating role of households. Household waste characteristics are being studied with a focus on per capita daily waste generation, and the percentages of fractions making up the waste (Figure 5.1). Besides the composition, also the factors affecting per capita daily generation and the composition of household waste are being discussed.

Waste characterization as defined by Mwai *et al.* (2008) is the quantification of various waste components. According to Dahlen (2005) the output of waste characterization is the weight and the composition of the various waste fractions. Data from waste characterization are essential for waste disposal facilities planning and waste management policy formulation (Chung and Poon, 2001). Bolaane and Ali (2004) attributed that knowing the waste characteristics is important to waste policy making and monitoring. In order to determine the per capita per day generation and

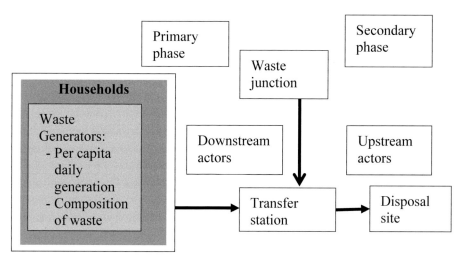

Figure 5.1. Households as waste generators in domestic solid waste management chains.

the percentages of the various waste components of household wastes in the selected study areas, we carried out a waste characterization study along the lines as suggested in the literature. There are three main methods for determining the composition of urban waste streams (Tchobanoglous *et al.*, 1993). The methods are:

- waste product analysis;
- market product analysis;
- direct sampling and analysis.

The methodology used in this study to determine waste characteristics was the direct waste analysis. The direct sampling method for determining the composition of the solid waste stream involves sampling, sorting, and weighing the individual components of the waste stream (Tchobanoglous *et al.*, 1993). According to Bandara *et al.* (2007), the direct waste analysis, although time consuming and labour intensive, provides reliable data that is detailed, accurate and informative when combined with factors affecting the waste generation. Because of the advantages of this method, we decided to use it to carry out the waste characterization study.

Qu *et al.* (2009) and Magrinho *et al.* (2006) indicated that compositional studies are important for several reasons, such as the need to estimate material recovery potential, to identify sources for component generation, to facilitate the design of processing equipment, and to maintain compliance with national laws. In addition, Al-Khatib *et al.* (2010), pointed out that the composition of solid waste is an important issue in waste management as it affects the density of the waste, the proposed methodology of disposal. For example, if solid waste generated at household level consists of large portions of kitchen or food waste, this indicates that frequent collection is needed due to its nature of decomposing rapidly and bringing foul smell. Information on the composition of solid waste is important in evaluating equipment needs, systems, and management programs and plans.

The waste generation is usually represented by waste generation rate, the quantity of waste generated per person per day (kg/day/capita). According to Bandara *et al.* (2007) the per capita waste generation rate is needed to predict future waste generation rates and for evaluating the waste generation trends in given communities.

Regarding the factors that influence per capita daily waste generation and composition, several studies have shown that a relationship exists between per capita waste generation and household size. The studies of Parizeau *et al.* (2006), Bandara *et al.* (2007) and Ojeda-Benitez *et al.* (2008) concluded that as the number of household members increases, waste generation per capita has been found to decrease. Bandara *et al.* (2007) also concluded that as the number of people in a household increases, there is a reduction in the per capita waste generation rate, thereby establishing the fact that when waste generation parameters are considered, per household waste generation is as important as the per capita waste generation rate. Other recent studies with similar observations include Van Beukering *et al.* (1999); Sujauddin *et al.* (2008), Jenkins (1993), Mosler *et al.* (2006), and Qu *et al.* (2009). Abu Qdais *et al.* (1997) found a statistically significant but weak negative relationship between waste generation per capita and household size in Abu Dhabi. While Bolaane and Ali (2004) show that there is a poor relationship between the number of persons in a household and the waste generation rate. They concluded that the number of persons in a household has a minor but discernible influence on the household waste generation rate. However, all these studies used statistical analysis to relate per capital daily waste generation with household size and income as the only important factors influencing per capita solid waste generation. In the present study also the effect of lifestyle is linked with daily waste generation and the composition of household waste. Determining the per capita daily waste generation and the composition of the household linking with the effect of social factors, makes a recent contribution to the knowledge of understanding the role of households as waste generators.

Unfortunately, few studies in the published literature have attempted to systematically relate lifestyle related factors to household waste generation. Yusof *et al.* (2002) pointed out that the effect of lifestyle is on the amount of waste generated by households. Tadesse *et al.* (2008) found that dust/ash and organic matter constitute the major components of household wastes as a result of the day-to-day lifestyle and consumption behaviour of residents. Also the study of Sabarinay (1997) showed the effect of lifestyle on the amount of waste generated by households.

In Dar es Salaam and other capital cities of East Africa there is no recent waste characterization study which has been carried out to document the per capita daily waste generation and composition of household waste. The earlier studies which were performed to characterize domestic waste include Kaseva and Mbuligwe (2005), who calculated the average per capita generation rate of domestic solid waste as at 0.42 kg/cap/day, while 0.39 kg/capita/day was reported by Kaseva and Gupta (1996) for low income households. These studies used door-to-door collection methods for assessing residential waste generation per capita, although it was not specified whether commercial wastes from home businesses were also present in the residential waste stream.

A summary of the composition of household waste studies carried out by JICA (1997), Shengena (2002) and ERC (2004) is presented in Table 5.1. The table shows the percentage composition in low income household waste for the 1997, 2002 and 2004 surveys in Dar es Salaam.

As can be noted, there was a slight increase in the percentage of kitchen waste from (39.8%) observed by JICA (1997) to 42% indicated by Shengena (2000). The study by ERC (2004), showed the same percentage of kitchen waste (42%). In general, the composition of household waste

Table 5.1. Percentage (%) fraction of waste materials in household waste in Dar es Salaam in 1997, 2002 and 2004.

Type of waste	JICA (1997)	Shengena (2002)	ERC (2004)
Kitchen waste	39.8	42	42
Paper	3.3	1	7
Plastics	1.9	1	4
Metal	1.8	4	2
Glass	1.3	2	3
Residual waste	34.7	16	16

contains more kitchen waste and residual waste than other materials, which is a typical households waste characteristic in developing countries.

According to the information in Table 5.1, recyclables, i.e. paper, plastic and glass amongst others are increasingly yearly. Plastic increased from 1.9% in 1997 to 4% in 2004, the paper fraction from 1% to 7%, glass waste showed a steady increase as well, and metal waste materials increased in 2002, but surprisingly decreased in 2004. The high percentage of paper, plastics and glass can be explained from the fact that the use of the materials especially for packing has increased, and become a source of environmental degradation.

Such materials provide recycling opportunities, while at the same time, a high fraction of kitchen waste must be considered both in terms of the additional problems it presents (such as foul smell, flies, etc.), as well as the potential opportunities for income generation.

5.3 Methodology

This section describes the data collection methods employed to generate information for the current chapter. Chapter 4 describes the methodological approach of the study in more detail.

The main tool used in data collection was a waste characterization study specifically conducted at 360 households sampled from six sub-wards in Kinondoni municipality in Dar es Salaam city. A combination of purposive sampling, clustering and random sampling methods was used in selecting the study areas and the respondents. The sampling procedure is detailed in Chapter 4.6. The selection of the study areas was based on the population density and the informal characteristics of these areas. The selected sub-wards included: Hananasifu, Kilimahewa, Makangira, Ubungo-Kibangu, Makoka and Mwongozo. The first three sub-wards are categorized as low-income sub-wards, while the later three are categorized as middle-income ones. The categorization was based on the results of the national population census of 2002. The six sub-wards were selected in consultation with Dar es Salaam city and Kinondoni municipal health officers and the waste contractors providing collection and disposal services in these communities. Table 4.4 presents the names and the main characteristics of the selected sub-wards, whereas Table 4.5 presents the number of households sampled from each sub-ward.

A structured household survey and waste characterization study were combined and carried out at the same time over a period of 3 month from March to July 2008. Prior to the survey, a reconnaissance survey was undertaken and discussions were held with respective sub-ward leaders and waste contractors providing solid waste collection and disposal services in these sub-wards to assist in the selection of respondents. The waste characterization study was carried out by the research team (researcher and trained research assistants) to determine the per capita daily waste generation and the percentage fractions of household waste constituents. Table 4.6 showed a summary of the materials and procedures adopted in the waste characterization study. The questionnaire was administered to the selected respondents by the research team. The information on household size and income were obtained from the first part of the questionnaire. The information on these two variables was important for this chapter, as the variables were related with the per capita daily waste generation.

Prior to this field study, interviews had been carried out with Dar es Salaam City Council and Kinondoni health officers as explained in section 4.2.2. Direct observations were used to gather information so as to corroborate the data from the published documents and the findings from the survey. Direct observation was employed throughout the period of study. For the current chapter information from direct observation focused on the types of waste generated by households.

A detailed content analysis of the relevant reports and documents on solid waste management provided secondary data. These documents included extensive and diverse literature from various solid waste management researchers around the world who published their findings.

Descriptive statistics such as means, frequencies, and standard deviations were computed by the use of the Statistical Package for Social Sciences (SPSS version 16). A bivariate analysis using Pearson's coefficient (r) was used to find the correlation between household size and per capita waste generation, and also between income and per capita waste generation. Data from observations and interviews were further processed and edited. Data entry into computer software was manually cross-checked to ensure accuracy.

5.4 Waste characteristics

This section reports the key findings of the solid waste characterization study conducted among the selected study population in Kinondoni municipality. The objectives of the study were to estimate the per capita daily generation rate of household solid waste (HSW) and to determine the relative proportions (percentages) of the household waste composition. The following sub-sections present the obtained results.

5.4.1 Per capita waste generation

Table 5.2 gives the details of the per capita generation of solid waste in each studied sub-wards as obtained from the field work. The mean values of per capita waste generation and household size are shown for each sub-ward.

The mean values of the average per capital household generation in each of the six sub-wards is shown in Table 5.2. The per capita waste (per person per household) was calculated by dividing the total waste generated by the total number of people in a household. The average

Table 5.2. Descriptive statistics of solid waste generation per sub-ward in Kinondoni.

Sub-ward		Waste generation rate (kg/capita/day)	Average household size
Hananasifu (n=70)	Mean	0.40	7.8
Kilimahewa (n=67)	Mean	0.33	6.8
Makangira (n=43)	Mean	0.36	4.2
Makoka (n=66)	Mean	0.48	4.5
Mwongozo (n=72)	Mean	0.56	6.2
Ubungo-Kibangu (n=42)	Mean	0.52	3.9

per capita household waste generation varied from 0.33 to 0.56 kg per person per day. The sub-ward of Mwongozo had the highest per capita daily waste generation (0.56 kg/capita/day), while Kilimahewa had the lowest per capita daily waste generation (0.33 kg/capita/day) among the six studied sub-wards. From these findings, an average of 0.44 kilograms of solid waste generated per person per day was computed for the entire study area. Table 5.2 also indicates that little variation existed in the per capita daily waste generation between the different the sub-wards.

The mean household size was found to be 5.82 as calculated from average means of each sub-ward. Average means were obtained through statistical analysis. The information on household size was important when calculating the per capita waste generation. The value of households size obtained in this research is slightly larger than what was reported by National Bureau of Statistics (2002)[14].

5.4.2 Physical composition of waste

Table 5.3 shows the average percentages of solid waste constituents found in household waste in each selected studied areas, as obtained from the waste characterization study. Food waste makes up the largest fraction of household waste at 70.0% followed by paper, plastics and residual waste, all of them at 8.0%. Aluminium, glass and metals were found at relatively lower percentages – 3%, 1% and 1% respectively.

The domination of the fraction of food waste (70%) in household solid waste can be explained by the fact that in the study sub-wards most of the food items were unprocessed with high moisture content, bulky and therefore denser. The food waste consisted of food leftovers and by-products such as peels from vegetables like carrots, cabbages, and fruits such as oranges and watermelon. The reason for the high content of oranges and watermelon peels in the household waste composition is because these fruits were seasonal fruits during the period when this study was being undertaken. The presence of these waste materials is the implication of its greater consumption by households when this research was being undertaken.

---

[14] According Tanzanian Bureau of Statistics (2002) the average value of household size in low-income areas was calculated at 4.9.

Table 5.3. Physical composition of household solid waste generated in studied sub-wards.

Sub-ward name	Waste category (%)							
	Kitchen/ food waste	Paper	Plastic	Glass	Metal	Aluminium	Residual waste	Total
Hananasifu	67	10	13	1.2	1.0	2	6	100
Makangira	60	17	9	0	0	3	11	100
Kilimahewa	67	5	7	3	5	2	11	100
Makoka	78	5	5	0	1	8	3	100
Mwongozo	80	5	7	0.5	1	0.5	6	100
Ubungo-Kibangu	69	7	8	1	2	2	11	100
Average	70	8	8	1	1	3	8	100

As waste comprised a high food waste fraction which was organic in nature, there are two main reflections: First and foremost this is an indication of composting possibilities, as the transformation of this material into manure (compost) could reduce the volume of waste required to be transported to the dumpsite. Secondly, when this kind of waste was not collected, there regularly emanated a foul smell due to its high decomposition rate, taking into consideration that the study sub-wards are found in a very hot climate[15]. When organic waste remains uncollected, it poses three major environmental threats: firstly, ground and surface water pollution through leachates; secondly, spread of disease vectors from open and uncovered waste dumped in crude dump sites; and thirdly, emission of methane which is a major greenhouse gas due to anaerobic decomposition in the dump sites[16].

Possible reasons for the higher amounts of papers and plastics might be twofold: first, affluence and increased use of plastics and papers as packing materials in retail shops and from business conducted within the household environment. Secondly, from our waste characterization study conducted, these materials were sorted already contaminated with particles and moisture from food waste which could not be fully separated from the plastics and papers. These adhered particles and moisture might be the factors contributing to the weight and possibly affect our waste per capita values. Our observations on households revealed that it is very common in Dar es Salaam to see people purchasing local food such as roasted cassava, local doughnuts (*maandazi*), and other items from retail shops wrapped in papers and plastics. Thin blue and black plastic bags are on the increase used as packing materials despite the governments' restriction of using thin plastics. It was common to see that after use people discard these materials with other waste materials in the same solid waste storage container. Compared to the findings of previous studies in Dar es

---

[15] The mean daily temperature for Dar es Salaam city is about 26 °C.

[16] For composting at household level see Chapter 6.4.3.

Salaam, the proportions of plastics and papers have been increasing (see Tables 5.1 and 5.3)[17]. In addition, these findings are relevant when comparing them to the guidelines for low income countries. According to the World Bank (Cointreau, 1982), low income countries should have a household waste composition by weight of 40-85% organic, 1-10% paper and glass, 1-5% metals, plastics, rubber, textiles and 1% residual waste.

An interview conducted with the Acting General of National Environmental Management Authority (NEMC), revealed that the Government of Tanzania has placed a ban since 2006 on local production of thin-film plastics used for making flimsy plastic bags in order to halt the further degradation to the environment caused by the rapid spread of used plastic bag litter. 'The ban of plastic bags and containers is necessary to protect Tanzania's rapidly degrading environment', vice-president Mr Ali Mohamed Shein[18] reported on 4th April 2006 (Pflanz, 2006) in The Daily Telegraph East African newspaper.

However, the NEMC official admitted that despite the ban imposed by the government in some places banned plastic bags are still widely in use. At the same time he appreciated that, some people have stopped using the bags and gone for other options, mainly used newspapers and magazines as well as paper bags that are believed to be more disposable and therefore friendlier to the environment. On my observation, I noticed that in the studied areas, many small shops and different types of business/commercial services were found within household premises. Therefore, it was difficult to distinguish between waste from domestic activities and that from business/commercial activities. There is no doubt that waste from these business activities adds substantial amounts to the total quantity of waste collected from the households.

It was noted that, households and shop retailers use the re-used magazines, newspapers available from homes and institutions (schools, institutes, offices) as wrapping materials. The householders use these papers to wrap food stuff and other items they sell. From the authors' experience, there were quite a number of people passing through residential areas and other places looking for old newspapers which they either get free or buy at a small price[19] to re-sell them to householders with small businesses. They pass around singing loudly: 'Magazeti – Magazeti' (meaning newspaper-newspaper) to alert whoever wants to dispose or sell to them used newspapers, papers or magazines.

Although we could observe that part of the waste was not strictly generated from conventional household activities, it is worth mentioning here that our results provide accurate and important information on the types of solid waste generated by households from the studied areas and can be used by policy makers for future planning. For example, the higher proportion of plastics and papers increases the volume of total waste and makes it more difficult to manage. Adequate figures on these waste fractions also enable more precise calculations on the benefits of recycling or (also informal) separation schemes and activities.

---

[17] Changes in the composition of waste between 1997 and 2008 showed an increase in plastics and papers which can be recycled and/or re-used. It suggests that the recycling potential tends to become higher.

[18] Tanzania vice-president from 2001 to 2010.

[19] Range between TZS 100 to 300 depending on the bulkiness. In 2008 when the fieldwork was conducted, $1 was equivalent to about TZS 1,330. In 2011, 1$ was equivalent to about TZS 1,545.

Despite the high presence of plastics and papers in household solid waste found in this study, there is need for further specification when considering the potential for recycling because the weighing exercise in our waste characterization study did not differentiate between recyclable and non-recyclable plastics. For instance, if plastics are grouped into only one category, their value is lower than when it is further separated into sub-categories of hard and soft plastics, etc. so as to meet the user's quality specifications. Therefore, in order to determine more accurate data for the recycling potential of household waste, another study should establish a further classification of the material composition of each category. It should be remembered, however, that, plastics are a major nuisance in household solid waste, because it litters the environment, clogs drains and causes flooding in the rainy season.

Other waste fractions such as metal, glass and aluminium were very small; this could perhaps be due to social and economic factors. The low presence of metal and glass in the household waste is the consequence of the trend that in recent times plastic materials have turned up as food containers, water bottles, medicine bottles, etc. where previously metal or glass containers were commonly used. In addition, according to information from the household survey, households recover plastic and glass containers by putting them aside for re-use, sale or to be given away as gifts. Plastic materials which were reported to be normally recovered include mineral water bottles. As indicated in Chapter 6, on average 71% of the respondents kept the recyclables for their own use, 7% indicated that they give them to others, and 22% mixed them with other waste in storage containers. Therefore, it is worth to note that these materials were recovered by households and not discarded as waste. What explains the low content of the aluminium waste fraction in the waste composition is that households do not normally treat used aluminium and other metal products as waste to be disposed of, since they can be sold or given to solid waste pickers.

Although the respondents were requested to keep the recyclable waste materials during the survey period, a low content of these recyclables was observed in the samples. Perhaps households kept these materials because they know their value and are not likely to include them in the survey, as other households prefer storing metal waste for later sale to scrap dealers. Currently, the demand for recycling metal scraps in Dar es Salaam is very high. It was common to see young people going around the city and within residential areas to collect scrap metals and aluminium. Middlemen buy the collected material from them and resell it to manufacturers for profit. Another possible reason which could explain the low content of aluminium is the low consumption of canned food, fruits and vegetables by households. Earlier, we mentioned that households in the studied areas tend to consume fresh and unprocessed foods which might surpass the consumption of canned products. We can conclude that, in fact, the percentage of aluminium, metal and glass determined in this study is insignificant.

The findings of the waste characterization study, revealed a relatively high amount of residual waste in the household solid waste composition. The residual waste contained a high content of charcoal ash, silt, sand residues and pieces of wood. It was noted during the survey and the observations conducted under this study that, in the daily household routine, the floor of the houses and yards were swept in the morning and in due course the sweepings were kept in containers holding other types of waste. According to the survey, 76.5% of the respondents reported that they mix sweepings from inside and outside the house with their other waste (see also Chapter 6). In addition to the reasons mentioned, the observations indicated that the overdependence on firewood and charcoal as a source of energy by the majority of households leads to the excessive presence of ash in household waste. It is worth mentioning here that the majority of households in the study areas and other parts of Dar es have the habit of using charcoal and wood for cooking. For instance, many people believe that rice cooks slowly and perfectly in a charcoal stove which is therefore preferred above an electric or gas cooker. This holds even for people who can afford other types of cooking fuel such as electricity or gas. In addition to this, the proportion of ash, silt, and sand may be high also due to the presence of unsurfaced yards and streets within the settlements. Furthermore, it was common to see in Kinondoni some people using paper and thin plastics as an igniting agent for their charcoal stoves, which may also increase the presence of ash in the solid waste composition.

The waste characterization study has provided useful data on the physical composition of household waste in the areas under study. Seven major components of the household waste have been presented with their relative percentages. These components are: kitchen waste, plastics, papers, metal, glass, aluminium and residual waste. Observation indicated that the composition of waste depends on many factors, such as waste handling practices when setting aside some of the waste materials, changes of material use and the circumstances of generating waste. These circumstances range from seasonality to cooking habits. In the following section we will explain the different demographic/socio-economic factors in relation to the different components in household waste composition in more detail.

5.5 Factors that affect solid waste characteristics

In our study, we identified different underlying factors which have a direct influence on the per capita waste generation and on the composition of household wastes. We classified these factors into two categories, namely demographic/socio-economic factors and lifestyle-related factors and activities. We tested the relationship between demographic/socio-economic factors and per capita daily waste generation using statistical analysis, whereas, the influence of lifestyle-related factors on per capita daily waste generation was judged on the basis of the findings from direct observations. The following subsections discuss how these factors affect the per capita waste generation and composition as revealed in our field work.

5.5.1 Demographic/socio-economic factors

It is important to remember that demographic/socio-economic factors are those related to personal characteristics such as income level, household size, age, gender, social class, level of education. However, our study considered only two of these factors, namely household size and household income to relate them with per capita daily waste generation. The reason for considering only these two factors is that they have widely been acknowledged as important factors influencing solid waste characteristics. But also, as explained in Chapter 3.4, a household consists of members that share their resources, whereas, household members pursue economic and social activities within household premises to provide for their daily needs and well-being, thus, it was important to include the household size and the household income in the waste characterization study.

Household size

Household size in the context of this study is the total number of people living in the same household. The household size for the studied households was captured during the questionnaire survey. As indicated in section 5.3.1, the mean household size was found to be 5.82. This figure is an average obtained by adding up the total of the average values of household size of each studied sub-ward and dividing this figure by the total number of sub-wards studied. It is important to mention that the majority of the respondents claimed that the size of their household was not constant and most of the time varying due to the extended families making up the households. Extended families mainly include incoming migrants from upcountry who can stay for a while, relatives visiting from rural areas, and others who may be in transit travelling to other parts of the country.

As mentioned earlier, the household size was an important parameter in this research since it was used to calculate the per capita daily waste generation. On the other hand, household size is one of the data needed in obtaining per capita waste generation, because traditionally the daily per capita waste generation is the ratio of daily waste generation to household size.

A graph was constructed using SPSS computer software to indicate the relationship between per capita daily waste generation and household size (Figure 5.2). Figure 5.2 shows the bar chart with error bars of the mean per capita daily waste generation per different category of household size. The results presented in Figure 5.2 show that for households with 1 to 4 members there is a direct relationship between size and the per capita daily waste generated, i.e. there is a constant increase in per capita waste generated with the increased size of households. The lag between households with members between 5 to 11 shows that for household with more than four persons

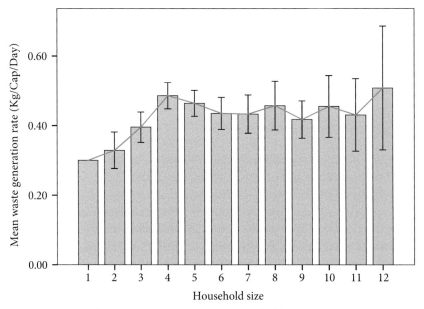

Figure 5.2. Per capita daily waste generation rate vs. household size.

the per capita daily waste generation is somewhat similar, it seems there is no direct relationship between these factors. Only for households with 11 to 12 there is again a direct relationship between the per capita daily waste generated and the household size. The sharp increase of per capita daily waste generation for households 1 to 4 and household size 11 to 12, shows that for small and large households the relationship between per capita daily waste generation and the size of household is stronger than for the middle range size 5 to 10.

Also a statistical method of bivariate analysis was used to test the correlation between the per capita waste generation and household size. A Pearson's coefficient of 0.04 was found. The aim was to find out if there is a direct effect of household size on per capita waste generation. As the value of 0.04 is very close to zero, this indicates there is very little or no correlation between the two variables. In another words, there is very little effect of households size on per capita daily waste generation in our study sample. Hence, the hypothesis that there is no relation between household size and per capita daily waste generated is accepted.

One would expect to find that waste generation on a per capita basis is inversely related to household size. This means that when household size increases daily per capita waste generation decreases (Jenkins, 1993; Ojeda-Benitez *et al.*, 2008; Qu *et al.*, 2009). So the larger the household size, the smaller the daily per capita waste generation. However, this is not confirmed in our findings. The results of our study indicate that the relationship between per capita waste generation and household size is not clear-cut. The possible reasoning might be as follows: the household size may not have an effect on the per capita waste generation, because, as our observation revealed, waste from business activities taking place at households are mixed with the waste produced from domestic activities. Therefore, the household waste production resulted not exclusively from domestic activities. Contributions to the household waste stream from business activities and other residue waste such as sweepings are independent of household size. As mentioned above, the per capita waste generation was calculated based on the values of household size as obtained in the survey. Another important observation as earlier mentioned was that the number of people living in a household could vary from day to day as relatives and friends move in and out. In this case, our waste per capita estimates may not be completely accurate because the sorting exercise in the waste characterization study was performed at three different times in a week for each selected household in order to obtain the average value. Whereas, the household size was recorded only on the first day of the sorting exercise. Therefore, in case the household size showed variations, this was not taken into consideration. Another possible explanation could be that from a statistical point of view the accuracy of determining these parameters increases with an increase in the number of samples that are analyzed.

Income

In this context the variable of income refers to the total household earnings from either informal or formal employment. Those who earn their income through informal employment are often self-employed. Examples of informal income generating activities which the researcher observed included amongst others roasting cassava and doughnuts, selling food, small shops, sewing and tailoring. As previously described these activities eventually contribute to the amount of solid waste generated by the household. Those formally employed receive regular monthly salary by being engaged either by the government or in the private sector. Income is measured in Tanzanian shillings (TZS) per household per month. As mentioned in Chapter 4, according to the respondents regarding to income earnings, 23 (6.4%) earn between TZS 0 to 50,000, and 117 (32.5%) earn between TZS 50,001 to 100,000, 60 (16.75%) earn between TZS 100,001 to 150,000, while the majority 150 (41.55%) of the respondents earn more than TZS 150,000 to 200,000 per month, and the remaining 10 (2.8%) more than TZS 200,000. It is important to remember that, the income reported by the respondents was calculated on the basis of monthly expenditures such as monthly bills on basic items such as food, house rent, water and electricity bills. The source of income is for the majority (185 or 51.5%) informal employment, while 153 (or 42.5%) had formal employment as their source of income, 22 (6%) could not specify their source of income.

Figure 5.3 was constructed using SPSS computer software to indicate the relationship between per capita daily waste generation and the household income. Figure 5.3 shows that households with an income of TZS 150,001-200,000 per month had the highest per capita daily waste generation, followed by households with an income of TZS 100,001 to 150,000. From the graph we can observe that the per capita daily waste generated by households earning TZS 0 to 50,000 per month is similar to households with an income of TZS 50,001 to 100,000 per month. Households earning more than TZS 200,000 per month had relatively lower per capita waste daily generation than those earning TZS 150,001-200,000 per month. In Dar es Salaam, households with an income of TZS 150,001-200,000 per month belong to the high income category, households earning TZS 100,000-150,000 belong to the middle income category, those earning TZS 0 to 50,000 per month belong to very low income category, and households earning between TZS 50,000 to 100,000 per

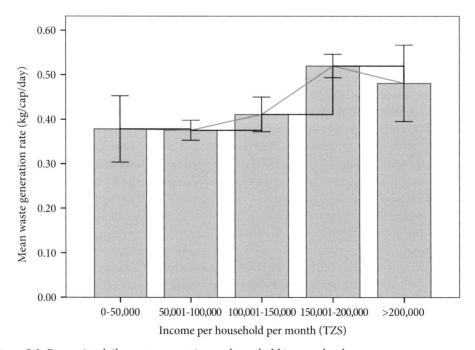

Figure 5.3. Per capita daily waste generation vs. household income levels.

month belong to the low income, while households earning more than TZS 200,000 per month belong to very high income class (NBS, 2002). This implies that households with high and middle income produce more household wastes (kg/cap/day) and therefore pollute more than the very low income households. This could indicate that they do more shopping and thus have more (super) market bags to throw away or have excess food which they throw away, e.g. after being stored in refrigerators, etc. The similarity of the per capita daily waste generation for households earning between TZS 0 to 50,000 and those earning TZS 0 to 100,000 may be caused by the similarity in lifestyle between these income groups in the areas under study.

As we can note, for the households earning more than TZS 200,000 per month, the trend is opposite, as can be noted from the graph there is sharp decrease of per capita waste generation. The per capita waste generation was found to decrease while the household income increased. The reasons for this surprising pattern may be contributed to the lifestyle, their living standard and the type of food consumed. Households with very high income may generate significant proportions of packaging waste, whereas, households with high and middle income generate wet food waste and significant amount of residual waste (sand, sweepings) which generally weighs more than the predominantly dry packing waste which may be common in households with very high income. It may be expected that households with very high income are more likely to buy packed foodstuffs and reading materials such as newspapers or magazines, than households with middle income who often prepare very basic meals. A study by Hockett *et al.* (1995) indicated that wealthier people often use and discard more paper, particularly as they read more newspapers. It is also possible that households with very high income are aware of environmental degradation and, therefore, have a culture of reusing or recovering waste, which would decrease the amount of waste generated. For example, recoverable materials such as plastic and glass containers are given to others, and food leftovers are kept in refrigerators and consumed at later times. Another possible reason of decreasing the per capita daily waste generated by households with very high income is possibly that members of these households dine outside (restaurants or hotels). This would reduce cooking activities at their households and lower the amount of waste produced by them.

The low-income and middle income households are more often involved in informal businesses.[20] These businesses serve as supplementary income generating activities which individuals working in the formal or informal sector employ as coping strategies to address the adverse effects from inflation or supplement their income.[21] The existence of informal businesses within household premises generates income and at the same time tends to contribute to more domestic waste in the same household. The incomes of those employed and at the same time engaged in the informal businesses tend to be higher than for those who solely depend on these businesses as their income generator. The low-income and middle income households are indeed on the margin and everyone tries to look for alternative sources of income. As previously mentioned, larger households have less income, and are thus getting involved in informal businesses to generate

---

[20] In the informal settlements people undertake development without a formal plan or approval. Thus, the development of business activities in these settlements follows the same trend.

[21] In 2008, the monthly legal minimum salary in Tanzania ranged from TZS 65,000 to 350,000. 1$ equals TZS 1,330. With this low payment from the formal and informal sector, informal businesses are not restricted to the category of the unemployed.

income and be able to sustain its members. Therefore, most of their domestic waste will tend to originate from the informal activities they undertake.

Households with very low income, have no reliable source of income and are therefore likely to generate less waste than those with regular sources of income. They may also have no or less household business activities, and thus generate less waste than those conducting such activities. It is true that a reliable income determines the purchasing power, such as the higher ability of the households to spend money on food. For example, Abu Qdais *et al.* (1997), showed that family income was positively related to the rate of household waste generated.

The statistical method of bivariate analysis was used to determine the Pearson correlation to measure the strength of a linear relationship between the per capita waste generated and the income of the households. The obtained value of the Pearson coefficient (r) indicates a positive correlation (r=0.379) between the per capita daily waste generation and the household income. This implies that the household's income contributes to the increase of the per capital waste generation. That is to say, the production of waste per person in households with relatively high income is higher than for those in the low income households. The hypothesis for this study stated that there is no relationship between per capita waste generation and household income. From our data of bivariate analysis we conclude that there is a relationship between these two variables and thus the null hypothesis should be rejected.

In our empirical investigations we found that social factors such as informal business activities and daily routines, also contribute to the generation of household waste. In the next section on lifestyle-related activities these factors will be discussed.

5.5.2 Lifestyle related activities

Given the context of this particular research, lifestyle refers to householders' daily social activities or the pattern of household daily practices that has an impact on waste generation at the household level. Lifestyle attributes which were seen to affect the waste composition in this research include daily cleaning routines, daily economic activities within households, recovery of recyclable materials, and the cooking and eating habits within the household.

Daily cleaning routine

Our study revealed that the daily cleaning routines of houses and yard, which take place in the morning hours, has an effect on the waste characteristics as householders mix sweeping-wastes from cleaning with other wastes in one container, thereby increasing the amount and weight of domestic wastes.

The findings revealed that solid waste management starts with general cleaning by sweeping inside and outside the houses. The general cleaning takes place before other domestic chores start. The survey and observations revealed that as far as the household daily routines are concerned, floors of houses and the yard around the households are swept in the morning, and the sweepings are discarded with other types of waste in the same containers. The survey indicated that 39% of households carry out cleaning as daily routine work between 5.00-6.00 a.m. and 50% between 7.00-10.00 a.m., while 11% of the respondents indicated that the cleaning takes place at any time during the day. The overall picture one can get from these findings is that the households' daily routines or social activities – here in particular daily cleaning within households – have very close linkage with solid waste management; therefore this link should not be ignored when proposing solutions for the SWM improvement at household level.

Daily economic activities

It is worth noting that lifestyle related activities also encompass daily economic activities. As explained in section 5.3.2, waste from businesses conducted by households influences the generation and composition of household waste. The daily economic activities of householders which we noted to have an effect on waste generation and composition are: small businesses which are undertaken by households as their source of income, and the pattern of (re)using plastics and paper as packing materials. From the observations made by the author, for many households there exists no clear distinction between domestic activities and economic activities.[22] The same observation was reported by Kachenje (2005). Furthermore, Kachenje observed that some of these activities can easily be accommodated into the domestic routine and thus contribute much to the economic status of the households.

In this study, seven out of ten households observed by author in Hananasifu sub-ward, undertook food-related informal business activities. Our survey results showed that among the economic activities, food-related businesses in the household takes the highest dominance (52%). These activities include preparation and selling of fish, fruits, vegetables, and *mama ntilie*.[23] Other economic activities are having small shops, selling charcoal, repair work (shoes, clocks, electrical facilities) and tailoring. These other economic activities altogether accounted for 26% of all households. Our survey results found that 22% of the respondents reported that they are not engaged in any household economic or business activities. There is no recent data which could be obtained to compare the findings of our study. A study by JICA (1997) estimated that 20% of the domestic waste was generated from the micro-businesses activities operated by family within its home. JICA stated that the main generators of waste were fruits, vegetables, fish and cooked food businesses. According to JICA, operators of these businesses constituted 67.5% of the total home-based operators. This implies that these economic activities play a vital role in the solid waste generation in the household in the areas under study and hence influencing the composition of household waste. Our survey indicated that out of 360 households studied, 281 (78%) of the total studied household operate informal businesses within their household environment. Based upon the high number of households involved in these activities it is expected that waste from these businesses attribute to increasing the amount and composition of domestic wastes.

With regard to the pattern of (re)using plastics and paper as packing materials was also noted an effect on the generation and composition of household waste. More goods made from non-biodegradable materials are now commonly in use, not just among relatively prosperous groups but in middle and low income households as well. Plastic bags and bottles, and packaging materials not only increase the volume of waste produced but also alter its composition, making disposal

---

[22] We classified economic activities into two groups as: (1) food related activities and (2) other activities.

[23] *Mama Ntilie* is Swahili word for female food vendor (s).

more difficult. This is compounded by the habits of people, where it is common practice to throw waste into the street, in open drains, on empty plots of land, or to simply burn it in the open air.

Recovery of recyclable materials

The recovery of recyclables materials is another observation on this particular subject. Respondents of the survey (see Chapter 6.4.2) confirm that they retain recyclables for their own use or give them away to others. It is evident therefore that, these kinds of waste materials were not included and counted in our waste characterization study. This is to say, the findings do not reflect the exact amounts of wastes generated at the household level, because the retained materials did not appear in waste stream.

Cooking and eating habits

Another lifestyle related activity which affects household waste generation is the intensive use of charcoal and firewood as cooking fuel and the specific eating habits. As previously illustrated, charcoal and firewood increases the content of residual waste since the ash content and the remains of charcoal are discarded together in one container together with other wastes. In addition, daily observations during the field work revealed that it is common practice in Dar es Salaam for people to use paper and thin plastics as an igniting agent for their charcoal stoves. It is inevitable that this contributes to the abundance of ash that is generated. The study also found that the specific lifestyle with respect to food consumption affects waste generation and composition. From the personal observations I realized that the high proportion of kitchen/food waste was found to be contributed by the habit of eating unprocessed food such as vegetables and fruits, and the low content of aluminium waste materials was, among other reasons, the result of not eating canned products. Further, we could note that a seasonal factor interacted with food consumption. This study however did not look at the seasonal variations like weekly or monthly differences in waste generation. As previously mentioned, this study was carried out in the period between May 2008 and July 2008 which coincides with the harvest period of water melon and oranges in Dar es Salaam. This impacts waste generation by having an effect on the amount of kitchen waste generated

It is worth emphasizing as a conclusion that social and lifestyle related aspects have a strong influence on waste volumes, composition and also waste management practices of households. Getting to know these social factors in more detail can be of great importance for future waste management policies and services.

5.6 Conclusion and discussion

This section discusses the main conclusions with respect to the research questions dealing with households as generators of solid wastes. The main findings are compared with the findings of other, related studies in the field, while key factors relevant for future policy making in this area are emphasized.

When confronting the main research question regarding the characterization of domestic wastes in Dar es Salaam, we estimated the per capita daily waste generation and determined the percentage of different waste fractions in domestic wastes composition. The average daily per capita household waste generation was 0.44 kg/per/day. These findings seem to be comparable with previous studies, although no recent waste characterization studies have been carried out in Dar es Salaam on this topic. The study by Kaseva and Mbuligwe (2005) calculated the average per capita generation rate of domestic solid wastes to be 0.42 kg/cap/day, whereas Kaseva and Gupta (1996) reported a mean of 0.39 kg/capita/day. In Uganda, Mugagga (2006) carried out a study in Makyinde division, where he revealed a mean of 0.55 kilograms of solid waste per person per day for middle and low income groups.

Our analysis of the waste composition showed the importance of looking at different components in domestic wastes and getting to know the reasons behind a specific composition. Kitchen wastes accounted for the highest proportion (approximately 70%), followed by paper and plastics (both accounting for 8%), while residual waste accounted for 8%. Aluminium, metals and glass all showed much lower percentages (all around 1%).

When discussing solid waste management from technical perspective, these results suggest the feasibility of recycling kitchen/food waste to ensure environmental protection and sustainable household waste management, because food waste is organic in nature and compostable in different ways, with the help of different technologies and at different levels of scale, including the local scale[24]. Also paper and plastics are recyclable materials and their presence in domestic wastes suggests again to take a closer look at the possibilities for re-use and recycling. Aluminium and metal are also recyclable materials but their presence is almost negligible. The analysis on paper and plastics offered in this chapter shows that informal activities and actors play an important role in solid waste management during the primary phase of the waste chain. When paper and plastics and other potentially useful fractions of the domestic wastes are at stake, the socio-economic relevance for local actors in the primary phase of the waste management chain needs to be taken into consideration.

When confronting the research question on the social factors behind the domestic wastes, this chapter discussed domestic solid waste management practices from a socio-economic and socio-cultural perspective. The results reflect the crucial role of the consumption patterns of households and the daily routine activities taking place at household level. Relevant consumption patterns refer to the cooking and eating habits of householders and their families, and they refer as well to the kind of materials or products used by households both for domestic routines and for home-based commercial activities. The food wastes generated by Dar es Salaam households originated not only from the normal preparation of these households' own meals, but also from food related businesses which are conducted at household level. Economic activities of households play an important role in shaping the volume and composition of the wastes generated. The residual waste fraction tells a social story as well, since it results from domestic routines of yard sweeping and from the use of charcoal and firewood for cooking. Understanding the dynamic composition of domestic wastes means getting to know the social dynamics behind their changing composition.

---

[24] Although the findings from the waste characterization study may suggest to see composting as an appropriate option for managing the study area's waste, results from the household survey raised questions about the feasibility of waste separation at source.

In the analysis of the social factors behind the volume and composition of domestic wastes, a distinction was made between demographic/socio-economic factors and variables on the one hand and lifestyle factors and activities on the other. Both kind of factors were shown to affect the amount and composition of per capita household waste generation in Dar es Salaam. Further statistical analysis showed that, household size and income level are not the key determining factors for understanding wastes generated by households as suggested in the literature. This chapter showed other variables to be important as well. For example, waste from sweepings and waste generated by business activities carried out within the household turned out to be important as well. In conventional analyses focusing on demographic factors only, these important variables are not taken into account properly.

It is therefore reasonable to conclude that future research on domestic waste generation should emphasize the daily activities and routines of households with respect to solid wastes instead of merely trying to link waste generation to income levels and household size. This is expected to result in more comprehensive data in the future. It is important that lifestyle factors are studied independently and for their own sake, in order to indicate and assess their crucial influence on the variation, the composition and on the trends in solid waste generation by Dar es Salaam households in the different sub-wards.

Chapter 6.
Household waste handling

6.1 Introduction

The previous chapter discussed the household waste characteristics by looking at the per capita daily waste generation and the composition of household wastes. Once waste is generated at the household level, it has to be handled in a manner that facilitates easy and (also ecological) safe disposal. Waste handling routines of householders involve all the activities that are associated with the management of waste in the primary phase of the waste chains. It is what householders do with their wastes until they are placed in storage containers for collection or otherwise handed over to formal or informal waste handlers in the secondary phase of the waste chains (Tchobanoglous *et al.*, 1993). Handling wastes in the primary phase also encompasses the transport or movement of loaded baskets or containers to points of collection. Before being transported to collection points however, domestic wastes needs to be stored somewhere properly. Handling of domestic wastes in the primary phase starts from the moment the waste is generated by householders and ends at the moment the wastes are taken over by other actors operating in the primary and secondary phase on both an informal and formal basis.

This chapter discusses the solid waste handling practices by households, and the solid waste flow from households to the transfer station. A transfer point is an intermediate place where waste is deposited and stored before being transported to the final disposal point. Transfer stations are known by a number of names, including transfer stations, transfer facilities, communal bins, and communal facilities. This study will use the term transfer station throughout for the sake of consistency. Transfer stations serve as a link between primary waste collection and secondary collection, i.e. community's solid waste collection program and a final waste disposal facility.

The objective of this chapter is to increase our understanding on the dynamics of household in handling waste. This objective will be achieved by answering the following questions as formulated in the theoretical chapter:
1. What are the current waste management practices applied by households?
2. How does the domestic solid waste flow travel from the households to the transfer station?
3. What are the different strategies that households use in managing waste?
4. What are the roles of different household members in the practice of waste management?

The study on waste management practices by households will provide knowledge on the different approaches in the management of waste and identify gaps that need policy intervention. While our study on the flows of waste within the household and in between the households and the transfer points, we aim to generate knowledge about the ways in which waste handling activities in the primary phase of solid waste management do or do not link-in effectively with the waste handling activities of other actors in the secondary phase of solid waste management. In that way, the understanding of householders' roles in solid waste management will help to identify factors to be taken into consideration when developing more effective SWM systems in in the future.

The chapter is organized into six sections. Section 6.1 provided the introduction, the research questions relevant to this chapter. Section 6.2 present and discuss the conceptual framework guiding the study. Section 6.3 deals with the methodology of the study. Section 6.4 concerns with the waste management practices at household level and section 6.5 analyzes the solid waste flows travelling from households to the transfer station. Section 6.6 describes the alternative methods of waste disposal used by households. Section 6.7 addresses the roles of different household members in solid waste management activities. Section 6.8 presents the conclusion and discussion of the main results.

6.2 Conceptual framework

The conceptual framework developed in Chapter 3 presents the three roles of households in solid waste management chain i.e. as waste generators, waste handlers and waste service recipients. Chapter 5 was guided by the first part of the conceptual framework on household as waste generators, while the current chapter is guided by the middle part of the framework on household as waste handlers as explained above. Figure 6.1 below presents the main concepts which are considered in this chapter.

The main components of household as waste handlers include waste (a) management practices at household level, (b) the transport of waste flows to the transfer station, (c) alternative methods used by households in managing domestic waste and (d) the roles of household members in SWM activities. The literature concerning most of these aspects is scarce and non-existent in the selected studied areas. Most of the studies such as (Cointreau-Levine, 1994; Van de Klundert and Lardinois, 1995) look at solid waste management from the perspective of the community and municipality

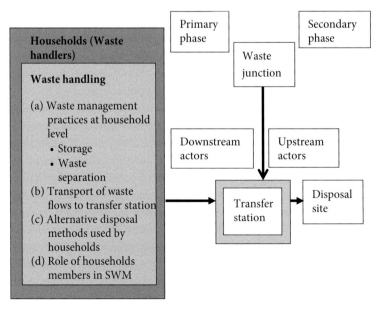

Figure 6.1. Households as waste handlers in solid waste management chain.

rather than the households which is the fundamental unit in the system (Addo-Yobo and Ali, 2003). Studies by Lardinois and van de Klundert (1993), Reem (2002), Tchobanoglous *et al.* (1993), highlight the problems of waste storage in hot climates in general with little focus on households.

Recent studies in East Africa on waste separation at household level include that of Okot-Okumu (2006) and Ekere *et al.* (2009), both carried out in Uganda. Okot-Okumu (2006) did not consider the practice of setting aside recyclables by households as waste separation, however, he acknowledged that components of wastes considered of value such as plastic bags and bottles for reusing are separated from the waste that is usually stored in a mix. Ekere et al discuss the reasons for households for not sorting waste.

On waste collection most studies regard it as a key link in the chain of SWM from the point of generation to the point of ultimate disposal (African Development Bank, 2002), and their focus is waste management at municipal level without paying attention to waste flows travelling from the households to the transfer station and linking these primary phase activities to actors, technologies and dynamics representative for the secondary phase of waste management. As pointed out by Dedehouanou (2004), the success of the transfer depends on the backward and forward linkages to waste technologies and infrastructures such as intermediate and final disposal sites, and the means of transport. He concluded that household waste collection schemes cannot be sustained without establishing strong linkages between the households and the municipality, e.g. in our terms between primary and secondary phase actors and factors.

Manga *et al.* (2008) indicated that in slums and unplanned settlements, indiscriminate disposal of municipal solid waste in streams, roadsides, vacant lots and low-lying areas are very common. However, there is no recent study which has documented on alternative methods used by households in managing solid wastes in the studied areas and in Dar es Salaam in general.

Anand (1999),discusses the alternative options taken by households for unreliable collection services. He established that one of the factor which determines the waste disposal or behaviour of households is the existing level of service. He highlighted that when primary waste collection services are not reliable the incentive is to explore other options and when regulations are either absent or the majority of peoples are non-compliant, the incentive is to dump wastes in open access spaces such as street and public spaces.

Regarding the specific roles of different household members in solid waste management, available studies report on role of household members at municipal level and on the general attitude of households towards waste recycling. Researchers as Chung and Poon (1995) and Nshimirimana (2004), reported that women are most involved in waste management, and that their attitude towards waste is decisive in the success of source separation of household waste and other recycling related activities.

According to Scheinberg *et al.* (1998) the roles of men are visible in waste management at municipal level where they are paid employees. While women are responsible for keeping the home and its immediate environment clean and disposal of waste is one of their daily tasks. In general women are not paid to handle wastes. They report that due to their restricted mobility and limited access to public spaces, some women, who cannot leave their homes for cultural or religious reasons, find it difficult to deliver wastes to a neighbourhood collection point.

Other studies on the role of household members include that of Ekere *et al.* (2009) who pointed out that women tend to be responsible for waste work within the household and along with children and domestic workers, may be involved in the sorting out and selling of recyclable materials. According to the study of Ali and Snel (1999), women are to a large extent responsible for household waste management, including the effective dealing with servants and informal waste pickers. Dedehouanou (2004) argued that in solid waste management, decision-making primarily lies in the hands of women. A quantitative analysis by Achankeng (2003) showed that children move as much as 80% of household wastes from the home to public bins.

6.3 Methodology

This section explains in brief the data collection methods employed in this part of the study. The details of methodological approach are given in chapter four. The main tool used in data collection for this chapter was the household questionnaire survey as it was specifically designed for the households sampled in this study. Prior to the survey research, interviews had been carried out with Dar es Salaam and City Council and Kinondoni health officers, while personal observation was an indispensible and important tool to gather information on the actual handling of waste in everyday life practices.

A structured questionnaire for households consisting of a mixture of closed and open-ended questions was used to gather quantitative information. The information gathered included: the households' solid waste management practices, waste flow to transfer station, alternative methods used by households in waste management and the roles of household member in solid waste management. Personal observation of households and transfer stations provided the opportunity to study the waste management practices within households in some detail, and to experience the practices of waste transfer from the households to the transfer stations. The field studies were carried out in six sub-wards namely Hananasifu, Kilimahewa, Makangira, Makoka, Mwongozo and Ubungo-Kibangu in Kinondoni municipality. These sub-wards were selected in consultation with Dar es Salaam City Council and Kinondoni Municipality Health officers. The selection of the sub-wards was based on the population density and the informal nature of the sub-wards. A total of 360 questionnaires were administered in these sub-wards and these served as the core of the empirical base of the study. The random sampling was employed to get the representative sample.

Hence, before the questionnaires were administered a reconnaissance survey was undertaken, and discussions were held with key informants to gain information about households in each sub-ward. Based on this information households were selected. The collected data through the questionnaire survey were analyzed, mainly with simple descriptive statistics; while for the qualitative data personal judgments, comments from experts, and results from interviews were used as a basis for the analysis and interpretation of the information.

6.4 Current waste management practices as applied by households

This section presents the findings concerning the waste management activities which take place after the generation of wastes at the household level. It includes the intra-household activities and routines which mainly comprise the storage and the (non) separation of domestic wastes

6.4.1 *Waste storage*

Household wastes are stored in different type of containers in the studied areas. The study showed the inadequacy of storage facilities. Despite, the official requirement from the Kinondoni municipality stated in Collection and Disposal of Refuse by-law, 2001 Section 4 (1 and 2) and Section 5 for all households to use a dustbin, a range of different waste receptacles including baskets, plastic containers, and boxes are used to store waste at the household level. Figure 6.2 shows the percentage of the most common type of waste storage containers used by households. The majority of the studied households use plastic bags locally known as *viroba* or *rambo* which are used to carry commodities home from shops and markets and later are used as waste storage facilities. As presented in Figure 6.2 of the total number (360) of interviewed households, 17.5% uses old plastic buckets, while 73.5% uses plastic bags with a volume ranging between 20 and 50 litres, 7% uses boxes of various sizes, and the remaining 2% do not have any container to store their waste. The old buckets in most cases cannot be used for any other purpose and containers such as boxes are dumped together with the waste.

Those without containers might be burning, burying the waste around their premises or taking away the waste to be dumped somewhere else. Due to the nature of the containers, households do not cover them, thus exposing the waste to flies, insects and rain. Moreover, households keep these containers outside the house. It was found that the majority of the households keep waste

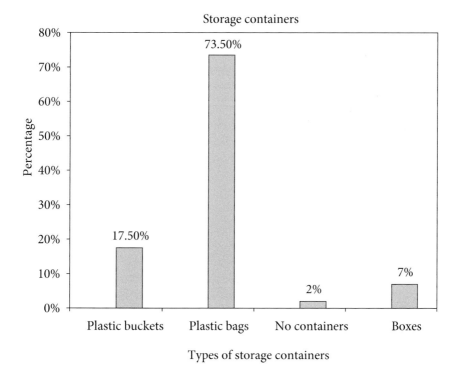

Figure 6.2. Types of storage containers used by households.

containers outside; because it is a cultural believe that it is not acceptable for wastes to stay inside. Elders used to say '*we cannot sleep with waste inside the house*'.

Among the problems mentioned regarding the practice of waste storage are inadequate space in the house to place waste containers, informal pickers retrieving the recyclable portion of waste and leaving the rest scattered all over the place, and the fact that the majority of the households – due to their low purchasing power – is not able to purchase and maintain the official standard solid waste bins. Some of households reported to have managed to purchase waste bins but they face the problems of theft by people passing by in the street. Waste storage bags are furthermore vandalized by domestic animals, especially dogs and cats, which tear them while looking for food; hence they spread the waste around the premises.

6.4.2 Waste separation at the household level

Waste separation at the household level was found not to be common practice in the households under study. Only two households out of the total number of respondents reported to separate kitchen waste (food waste, vegetable and fruit peelings) to feed pigs. Of the 360 respondents, none acknowledged that they separate their household wastes. This complicates solid waste recycling for other uses and composting of waste.

However, according to information from the survey respondents, some households recover plastic and glass containers by putting them aside for re-use, sale or gift. As shown in Figure 6.3 the findings indicated that on average 71% of the respondents kept the recyclables for their own use, 7% indicated that they give to others, and 22% mix them with other wastes in storage containers.

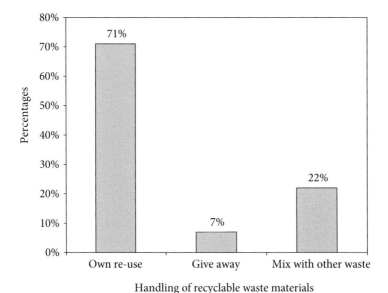

Figure 6.3. Handling of recyclable waste materials.

The most common recovered materials are plastic materials mainly mineral water bottles, and glass containers. Households consider these materials to have value for re-using and that is the main reason why they are separated from the rest of the waste materials. Some respondents reported that, they put plastic bottles and glass containers aside for waste pickers, to avoid scattering of waste which is caused by unknown illegal waste pickers who pass around retrieving recyclables from storage containers leaving other waste scattered around. From these reasons, we can say that setting aside these materials is not because households are aware of environmental issues regarding solid waste management. Furthermore, when respondents asked an open question as to why they do not separate waste, various reasons were given as indicated Table 6.1.

The results in table show that the respondents indicated reasons for not separating waste at source. The bigger percentage (43%) of respondents indicated that this practice it is not known to them. The second most frequently mentioned reason for non-separation is the time element (20%). They view it as wastage of time, because according to respondents even if they separate the waste, the waste collectors later tend to re-mix it. This is a fact because throughout the Dar es Salaam city solid waste is collected as mixed waste. Moreover, they might regard it as a wastage of time mainly because they have not seen any economic benefit out of it, and there is no emphasis on waste separation from the side of official policy. But also households can perceive waste separation as time consuming because they have not been practicing it. Additionally, 10% believed that waste separation was time consuming. The waste of time argument also played a role during the waste characterization exercise organized in the context of this PhD-research, when respondents failed to sort out waste into different fractions for weighing by claiming that it was time consuming. This means that even for those with knowledge on resource recovery, there are other factors at play which are making the implementation of waste separation policies difficult. It is also interesting to note that a smaller percentages of the respondents indicated other reasons, such as no space for containers (7%), the fact that they cannot afford to have more than one container, 2% is not interested in sorting the waste and 1.5% indicated that it is expensive to have more than one container. While 16.5% indicated that no need to separate waste when there is not much waste generated.

Table 6.1. Description of reasons for not separating solid wastes.

Reason for not separating waste	N	Percentage
Waste of time	72	20%
Time consuming	36	10%
Expensive to have more than 1 containers	6	1.5%
Not known to me	154	43%
No space for containers	25	7%
Not much waste for separation	59	16.5%
Not interested	8	2%
Total	360	100%

The amount of recoverable materials in the waste stream is limited because low-income households rarely use items that are packaged and rarely buy bottled or canned items which are the predominant source of recoverable materials. Moreover, households realize the cost of storage in waste separation. Most of them reported that it is easier and cheaper for them to collect the waste together than to separate it. They reported that they do not segregate the waste because it is time wasting for them, since the workers carry all the waste together in a lorry to the dump site.

6.4.3 Composting at the household level

As mentioned in section 6.4.2, of the 360 survey respondents, none acknowledged that they separate their household wastes. It is clear from our findings that one of the factor amongst others which will hinder composting is non-existence of separation of generated waste to remove compostable kitchen materials to effect composting. Our waste characterization study found that kitchen waste, a lot of which is compostable, contributed to 70% (see Table 5.3) of household waste and is higher than other solid waste materials generated at household level. Solid waste separation is one of the important elements of effective composting. Clearly, non-separation of the solid waste at source observed in this research reduce the chance of setting up composting schemes at household level. In addition, for the sub-wards under study composting at household level will be hindered by inadequate space as there is no land available for composting within these sub-wards any more; the land is congested and used for mainly residential purposes.

It is documented (Zerbock, 2003) that household-level composting has the greatest potential for success in many areas, especially those where small scale agriculture is found in great abundance close to urban areas, or those where limited gardens are found within the city itself. Currently, Mkwela and Banyani (2008) carried out a study in Dar es Salaam, and noted that urban farming practices in Dar es Salaam is often carried out in inefficient ways, with no access land, and no access to farmers' markets. This means that the key issue in composting is to find market for the final product, either by selling to neighbourhood farmers/gardeners or on the household's own plots. Mkwela and Banyani (2008) added that there is no financial or infrastructural support from the central and local government to promote farming in the city. In this case we, therefore, argue that, even if composting activities are initiated in the sub-wards under study they may fail due to the prevailing constraints. The most important advantage would be that composting would reduce significantly, the amount of waste that is currently being disposed off reducing the life of dumpsite.

6.5 Waste flows travelling from the household to the transfer station

Household waste flows travelling to the transfer station in the context of this study refer to the practices of collecting and transferring domestic solid wastes which is stored in the households to the transfer station which are at some distance from the household.

6.5.1 The formal collection systems

The formal systems for collecting waste flows from households are the ones carried out by waste contractors with legal contracts for providing solid waste management services from Kinondoni municipality. The study found out that formal systems were applied in the sub-wards of Hananasifu, Makangira and Kilimahewa. Whereas, in the sub-wards of Makoka, Mwongozo and Ubungo-Kibangu there was no formal system during the period of undertaking this research, and therefore households in these sub-wards used alternatives methods to remove waste from their premises. According to the findings of the study, the formal systems were involved in the transfer of domestic waste flows to the transfer stations. According to the interviewed waste contractors, the number of households which received formal solid waste collection was 2,430, 880 and 2,000 in Hananasifu, Kilimahewa and Makangira respectively.

The collection of domestic solid wastes was primarily carried out by solid waste contractors using two major methods: door-to-door collection services and 'brings systems' using waste trucks. In Hananasifu and Kilimahewa sub-wards waste contractors apply both methods of waste collection, while in Makangira sub-ward, the bring system of waste collection was being applied. The following section describes the application of the two methods of waste collection in the studied sub-wards.

Door-to-door method of waste collection

Door-to-door methods of waste collection were found to be practiced in Hananasifu and Kilimahewa sub-wards. Basic equipment such as wheelbarrows and pushcarts were used to collect the wastes at the door step of the households and to deliver the wastes to the collection point (or transfer station) in the form of standby trailers. Personal observation and interviews with waste collectors indicated that, door to door methods were provided to households living in areas where pushcarts and wheelbarrows can penetrate and access wastes from households. KIWODET reported that within Hananasifu sub-ward, in some areas where roads are passable, waste collectors use trailers pulled by tractors to collect waste through door to door method.

The use of handcart is very suitable for conditions prevailing in the low-income areas, such as narrow streets, low generation rates and low wages. Manually operated carts are cheap in manufacture and operation, quite simple in design, and can be produced and maintained locally, which are all important prerequisites for a sustainable technology. The main advantages of this method are that much waste is ensured to be collected and it is convenient to households. However, the method is time consuming and very costly especially if labour costs are high.

As revealed from the interviews with waste contractors, waste collectors or labourers are employed on a daily basis to collect waste from households. As such, the number of labourers employed per day by waste contractors, vary from one contractor to another but generally it ranged from 3 to 5 workers. According to KIWODET, they employed 3 to 5 labourers per day for collecting and transferring waste to the transfer station. These labourers were paid between TZS 1,500-3,500 per day (about USD 1.13-2.25). Tua Taka Makurumla (TTM) had employed 6 labourers whom were paid TZS 2,500 per day (about USD 1.88); According to CLN Electrical and General Contractors Ltd., they were not employing labourers of collecting waste from households, because households themselves bring waste and dump into their waste truck.

The bring system of waste collection

The second method used by waste contractors to collect waste from households is the bring system. As mentioned in section 6.5.1, this system was found to be applied in Kilimahewa, Makangira as well as Hananasifu sub-wards. With bring systems households were responsible for bringing their waste to areas where waste trucks or stand-by trailers were located. So as to ensure that potential locations are suitable where all households can reach easily to dump their waste the point to locate transfer station is decided upon by households, waste contractors and the local leader of the respective sub-wards. The information obtained from the interviews with the service providers as far as the furthest distance travelled by households to reach the transfer station ranged between 200 to 500 metres (see Table 6.2).

The bring system was mainly adopted in these sub-wards because the accessibility of households in the area is poor and paths between houses are so narrow, hilly and depressed lands to the extent that even wheelbarrows cannot pass. Waste from these inaccessible sub-wards was carried by household members largely by women, children and young boys, to an identified place where the cart, wheelbarrow or vehicle can reach. In Kilimahewa and Makangira sub-wards, vehicles stand at a certain agreed point on the street within the settlement for easy accessibility where household members bring their waste. These trucks have a system of horn or a sound to alert households to bring out their waste which is then emptied directly into the trucks by the crew members and the containers are returned to the owners. When a collection waste truck arrives in the area to be served, normally one of the crewmembers would walk along the area alerting the households to bring out their waste in containers. Waste in containers such as boxes and plastic bags were dumped together with the containers. When the trucks were filled by the contents brought, the waste was transported to the dumpsite. For the case of Hananasifu sub-ward, households living in inaccessible places are required to bring their waste to standby trailers which were stationed in the premises of the office of KIWODET. According to the interview with KIWODET spokes lady, the reason for choosing the premises of their office to locate the standby-trailer, was to enable them to control dumping of waste in a transfer station to ensure that waste was dumped by only their registered customers.

This system is cheap in terms of time required for collection; however, if household members are out when the waste truck arrives, waste must be left outside for collection which may be scattered by wind, animals and illegal waste pickers. The truck-system on the other hand appears to be the most efficient but can only be applicable where there is a possibility of accessing the area by a vehicle. It is noteworthy to know that these waste contractors also provide solid waste collection services to other sub-wards which were not included in this study.

Frequency of collection

According to waste contractors solid waste collection is done between 7.00 am and 12.00 pm, and in terms of waste collection frequency they reported that collection is between 2 and 3 days per week as shown in Table 6.2. However, according to our survey, when respondents were asked 'how many times per week is waste collected from the households', responses given were found to be different from those revealed by waste contractors. 60% of respondents indicated that waste is collected once per week, 12% indicated twice per week, 3% said waste is collected now and then, while 25% could not give any answer. Respondents complained that in several cases waste contractors fail to collect waste on the scheduled days. The uncollected waste is vandalized by animals such as dogs, leading to the littering of the surroundings and a bad smell of decomposed waste. In conclusion, Table 6.2, shows an overview of the details of the formal solid waste collection systems based on the interviews with sold waste contractors.

Inadequate waste collection was attributed by the contractors to frequent mechanical problems with their collection vehicles, traffic jams, the long distance to the dumpsite, and the inaccessibility of some settlements. Due to these problems, the scheduled services are not adequately delivered, and the delays occur. As said by KIWODET spokes lady:

> Given the condition of swampy and depression land, it is not possible to access and provide waste collection services in Bonde la Mkwajuni[25] sub-ward (KIWODET spokes lady Mrs Msosa, 8/10/2007).

Table 6.2. Waste collection details.

Sub-ward	Collector[1]	Method	TS[2]	Furthest distance to TS (km)	Equipment owned	Rate of collection	Time
Hananasifu	KWD	Door to door, bring system	Stand-by trailer	0.4 km	5 pushcarts; 1 trailer; hire trucks; 5 wheelbarrows; 1 tractor	2 times per week	8:30-12.00
Kilimahewa	TTM	Bring system	Waste truck	0.5 km	4 pushcarts; hire trucks; 1 truck; 2 trailers	2 times per week	8:00-15:00
Makangira	CLN	Bring system	Waste truck	0.2 km	1 tractor; 2 trailers; 1 truck	2 times per week	7:00-12:00

[1] CLN = CLN Electrical and General Contractors Ltd.; KWD = KIWODET; TTM = Tua Taka Makurumla.
[2] TS = Transfer Station.

---

[25] Settlement within Hananasifu sub-ward.

6.6 Alternative disposal methods used by households

This study found out that households in sub-wards with no formal system of waste collection are compelled to find other ways to get rid of their waste. The following section describes the alternative disposal methods that households use in managing their waste. These methods are referred to in terms of 'informal methods' because they are unauthorized methods.

6.6.1 Informal systems for waste handling

The informal systems in the context of this study refer to the other options besides using the formal waste collection system. Informal systems are used by households to remove domestic wastes from their premises. The informal systems identified in this study were dominated by practices such as burning, burying and illegal dumping. In addition, households also pay illegal waste pickers to remove the waste from their premises. In several cases these different informal practices were applied in combination.

During the period of this study (2008), formal waste contractors had withdrawn from providing collection services in the middle-income sub-wards (Ubungo-Kibangu, Mwongozo and Makoka – all located in one ward). Although, respondents were not aware of the reasons for the waste contractor to pull out of their area. When interviewing a Kinondoni SWM official, he reported that waste contractors failed to provide solid waste collection services to the assigned areas due to various reasons such as a failure to collect enough waste fees and the breakdown of equipment. He added that some of waste contractors entered in solid waste management business as a stepping stone, hence they do not invest and innovate much in it and rather re-invest their proceedings and develop interest in other businesses, e.g. hotels.

For not having formal waste contractor to provide waste collection services, households Ubungo-Kibangu, Mwongozo and Makoka were compelled to find alternatives to get rid of their wastes. However, it was also revealed that informal systems operate alongside the formal system in Hananasifu, Makangira and Kilimahewa. Figure 6.4 presents in summary the alternative waste management practices as applied by households.

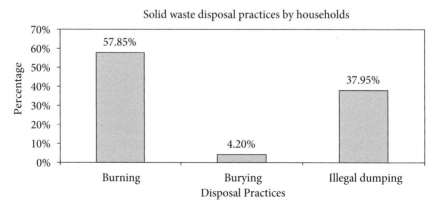

Figure 6.4. Alternative waste management practices by households.

Although the statistics regarding illegal pickers were not included in the subject of this study, relevant information regarding illegal waste pickers was obtained from the households through informal discussions and personal observation. According to the respondents, illegal pickers charged waste collection fees according to the volume of the wastes ranging between TZS 500-1000 per collection[26]. The payment is made on the spot. Interviewed households indicated that informal waste collection practices were more reliable than the formal collection services by waste contractors. On further discussions with interviewed households, they revealed that some of the illegal waste pickers have specific households which they make sure that other pickers are not allowed to serve. By having their specific households, they are sure about the source of income and know when to visit their customers (households) to collect solid waste.

When the respondents were asked whether they knew where these informal pickers disposed the collected waste, some mentioned open places and others in valleys, but the majority said they did not know where they disposed it. As one respondent said:

> Actually when they take our waste, we do not ask them where they take waste to. But as you can see there are many sacks of waste (viroba vya taka), scattered all over in our settlements may be they just dump anywhere provided they are not seen by anyone. But as s long as my waste is collected, I don't care about where that waste goes (Makoka sub-ward survey respondent, June 2008).

Another respondent said:

> We have directed them (waste pickers) to dump waste they collect in the valley adjacent to army area (jeshini), because these boys collect waste and sometimes they just dump on roads or in front of the house of a neighbour (Makoka sub-ward survey respondent, June 2008).

The interviewed solid waste contractors complained that illegal waste pickers dump the waste in the stand-by trailers stationed in sub-wards which they operate. As expressed by KIWODET interviewee:

> One of the big problem is that our customers (households) do not pay us because they give waste to illegal pickers, and these pickers transfer that waste into our transfer station, and they usually do it at night (KIWODET spokes lady Mrs Msosa, April 2008).

A large number of households burn their waste near their houses, on open sites and by roadsides. According to the respondents, bulky waste such as paper, thin plastics and old clothes are burnt to reduce the volume which consequently lowers the collection charges. Based on the findings, as indicated in Figure 6.4, the dominant waste disposal carried out by studied households is burning and illegal dumping. Burying waste is the least common method used. Only 4% of respondent households burned their waste, and 37.95% dump their waste illegally, while 57.85% burned their

---

[26] 1$ = 1,330 TZS.

waste. According to respondents, burning and illegal dumping are more favoured than burying because these practices are not expensive and easy to operate. As said by one respondent:

> *I ask my children to burn papers and old clothes to reduce the volume of my waste, because I don't have a big waste container (Survey respondents, April 2008).*

Another respondent claimed that:

> *We burn paper waste because it catches fire easily and burn very easily (Survey respondents, April 2008).*

A respondent from Hananasifu sub-ward claimed that the alternative is to burn our waste because there is not enough space for burying. When she was asked why she practices burning despite having formal waste collection, she claimed that waste contractor (*mkandarasi wa taka*) attends their area once or twice per month. This was an indication for the poor SWM services provided by the waste contractors to the studied households causing them to manage waste themselves through methods that are not environmentally safe. According to KIWODET (legally assigned to provide waste collection in this area) admitted that sometimes in some occasions the collection service experiences a problem with the collection crews so the scheduled service can be absent or the area is attended on a delayed schedule.

In areas that did not have formal service, some households practiced waste burning or burying their waste routinely because they have no other option. Figure 6.5 presents the different formal and informal methods used for domestic waste handling in the sub-wards under study.

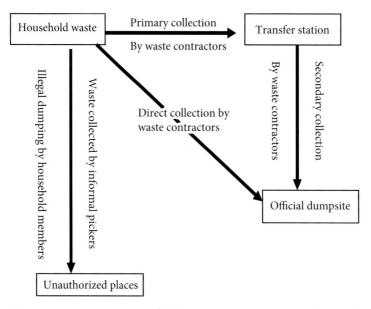

Figure 6.5. Waste management practices for domestic wastes in Dar es Salaam sub-wards.

6.7 Roles of different household members in the practice of waste management

This study revealed that the generated solid waste at household level was handled by different household members. Different household members play different roles in waste management practices. Through observation, it was apparent that women have the primary responsibility for cleaning, food preparation, and household maintenance, so women were also directly involved in solid waste generation and management.

Survey and observations of the daily household routines revealed that women, housemaids and children usually manage and dispose of the household waste as part of their domestic work. Broadly speaking, it was women, housemaids and children who had the responsibility for keeping the household surrounding clean, taking out the wastes, and transporting it to the transfer station. As one woman from Hananasifu sub-ward put it:

> *It is not usual for a father of the house to deal with waste. This is job for children or myself. Sometimes I ask money from my husband to pay tax for waste (ushuru wa taka) (Hananasifu sub-ward respondent, Mama Jamila, April 2008).*

This statement indicates that women also have to pay for solid waste collection fees. It is not always the case that the husband (man) has to pay.

This division of responsibilities was verified also in the household survey, which confirmed that in 74.5% of the households the housemaids/children are responsible for waste management, in 24% the mothers and in 1.5% other members of household as indicated in Figure 6.6.

We also noted that women were very creative to identify the recyclable material from wastes which they later re-used in their own households for different purposes. As discussed in Chapter 5, there was a practice of setting aside recyclable materials such as plastic bottles and glass containers.

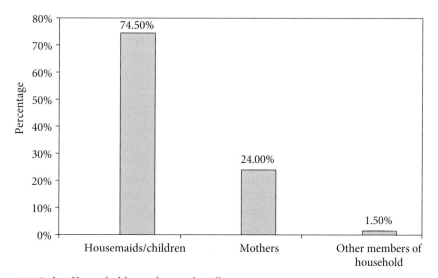

Figure 6.6. Role of household members in handling waste.

This practice was observed to be done by women. For example glass bottles and plastic bottles were washed and reused for storing cooking oil, salt, sugar, kerosene and local brew. In addition, the use of wastes as source of cooking fuel such as coconut husks and shells, thin plastic were observed to be common done by women. In ten of the studied households which were personally observed it was found that, in all of them, housemaids, and children were responsible to transfer waste to the transfer station. When the waste truck was arriving to collect the wastes, housemaids or children carrying waste out alert one another on the arrival of truck. It was also noted that it was a common practice to transfer waste of the neighbour. The male members of the households rarely participated in household waste management activities, except that they sometime involve themselves when the waste was bulky and some physical help was required to transport it away from the household to the transfer station. Other tasks mentioned with respect to involving men in waste practices were activities such as digging holes for burying waste, and the burning of waste. Another important observation was that of the 360 respondents, women respondents represented 77.5% and men only 22.5%. These findings shows an indication of the distinctive roles of women in waste handling within their households.

6.8 Conclusion and discussion

In an effort to confront main research question 2 this chapter explored the role of householders in the management of domestic wastes after it has been generated by the households. Different tasks or practices of waste-handling in this primary phase of the waste chain were explored with the help of both qualitative and quantitative research methods.

The overall picture that emerges from our empirical research confirms the widely recognized fact that there is a lot to be improved in the present situation. By looking in detail at the different tasks and responsibilities and by reporting also on the reasons behind the lack of proper implementation of the waste management tasks to be performed, we aim to gain a better understanding of the present situation and to identify the key factors relevant for improvements in the waste management practices of households in Dar es Salaam. We discussed in some detail the practices of storing wastes, waste separation, and ways of (non) handing over the wastes to both formal and informal collectors who connect households to the secondary phase of the waste chain.

With respect to the practice of storing domestic wastes, it turned out that official regulation was not implemented. Containers were used without lids, and were located outside in the yard of the houses for collection. The use of various types of storage containers was found to be contrary to the requirements of the existing municipal by-law which stipulates that each households should have two storage containers one for organic and the other for non-organic waste of not less than 40 l fitted with a lid. The common use of various types of waste containers clearly shows that the municipal SWM by-law is not adhered to by households.

There are many possible explanations for the distance between official regulation and the actual situation on the ground. Next to a lack of law enforcement and householders being ignorant about existing regulation, we found a number of more specific, situation bound factors which can explain existing storage practices of domestic wastes in informal settlements. The prevailing physical and social conditions do not allow household to use the municipally prescribed waste containers. As we noted in the survey and focus group discussion, inadequate space, theft of containers and lack of financial resources are among the most important factors to explain the existing situation. When householders are supposed to bear the costs of (official) storage containers, householders choose to use very old, broken containers that are used only one time and disposed away together with the wastes. Householders in informal settlements choose the kind of storage containers which do not have (economic) worth to others (to prevent theft), which are easy accessible (obtained in the shops) and which require no maintenance (cleaning). Plastic bags instead of official containers 'fit' in the situation on the ground and correspond better to the existing waste management practices by householders.

With respect to the practice of waste separation at source, the study revealed that, Kinondoni municipality currently has no policy or programs that encourage waste separation at household level. The study, however, showed that some households keep separate recyclable materials such as plastic and glass containers, and do not mix them with other everyday household wastes. This means that households do not regard these items as wastes, rather, they have realized that the waste materials they keep aside are still valuable materials to them and others. These householders practices of setting aside valuable materials is not recognized by or supported by official authorities, even though these practices contribute to reducing the overall costs of solid waste management.

When considering the reasons behind the lack of separation of organic wastes (with 70% a very important category of domestic solid wastes) the study found that householders in general do not show a clear understanding of what waste separation is about, while at the same time indicating that forms of waste separation as officially required by the formal regulation at present are very difficult to comply with, given the situation on the ground.

With respect to the practices of the (non) handing over of domestic wastes to both formal and informal collectors, the main conclusion is that there are many different practices existing side by side. There are formal collection schemes and also schemes in which householders bring their wastes to the waste transfer point; there are informal waste collectors who are more or less prominent and permanent, depending on the (lack of) performance of the official collection system; there are ways of getting rid of wastes in an 'alternative way', by having the waste burned, buried or otherwise. We explored in detail the responsibilities of transporting the wastes from the households to the transfer point and were able to show the significant role of women and children in this practices. We showed that the private sector is involved in primary waste collection by using handcarts and wheelbarrows to collect waste from inaccessible locations through door-to-door collection, and waste trucks to transport the wastes from the transfer point to the dumpsite. The successful use of handcarts and wheelbarrows for collecting waste in such settlements shows that these equipment are appropriate technologies, adapted to the physical circumstances in informal settlements The key role of informal waste collectors who collaborate closely with householders in the primary phase of the waste chain turns out to be of significant importance and needs to be understood in direct relation with the (lack of) performance from the side of the official, formal waste collection system. Because of their important but not formally recognized role in waste collection, we discuss in Chapter 7 the potential of given informal waste handlers a more officially recognized role in waste handling practices for example with the use of so called Public Private Partnership arrangements.

With respect to the practices of 'alternative' or 'illegal' handling of domestic wastes the main conclusion is that alternative ways of getting rid of wastes are particularly prominent when official systems do not perform well. Illegal practices may be considered first as an indication of poor solid waste services provided by the formal sector, and secondly, lack or poor law enforcement by the municipal authorities. Once the alternative (and cheap and unhealthy) ways of getting rid of domestic wastes are established and routinized, it turns out to be more difficult to replace these unsustainable practices by more sustainable ones.

Chapter 7.
Households as service recipients in solid waste management chain

7.1 Introduction

The previous two chapters studied households as generators of solid waste and as handlers of solid waste in the primary phase of the solid waste management chain. As earlier discussed, households generate various types of waste which include kitchen waste, papers, plastics, metals, aluminium, glass and residues such as ash and sweepings in different proportions. They were also involved in solid waste management though various activities, such as waste storage, collection and the transfer and disposal of domestic wastes.

Since households are waste generators, and also participate in waste management activities, they must be regarded as major stakeholders in solid waste management (Mosler *et al.*, 2006; Snel and Ali, 1999). This chapter aims to confront research question 3 by exploring firstly how householders as stakeholders in the SWM-chains receive solid waste management services and secondly what their basic perceptions and evaluations are with respect to the ways they are being served by other actors in the SWM-chain.

According to Joseph (2006) stakeholders are people and organizations having an interest in good waste management, and participating in activities that make this possible. They include enterprises, organizations, households and all others who are engaged in some waste management activity. Anschütz *et al.* (2004) describes other key stakeholders to include local authorities, NGOs/CBOs, service users, private informal sector, and private formal sector. Also from the literature, households are identified as the main and primary stakeholder especially in solid waste services at the local level while contributing in particular to the sustainability of the private sector in solid waste collection (Snel and Ali, 1999). However, as highlighted by Ahmed and Ali (2006) people are often overlooked in the service delivery framework, while they can contribute significantly to service delivery. But more importantly, they can play an active role in improving accountability and service quality of both public and private sector. Functional links between households, community-based activities and the municipal system are very important, even where municipal waste collection services are provided in a regular and formal way, the cooperation of householders is essential to efficient SWM operations.

To make solid waste management work at the local level, there is a need of gaining legitimacy and support from households, and to understand their perceptions towards solid waste management. When support and legitimacy are absent and perceptions negative, one can expect householders to resort to open burning of wastes, when materials are simply set on fire and left to burn. When support is lacking, we see householders not becoming involved in waste separation or householders not willing to deliver domestic wastes to a transfer station or communal facility. Considering these examples, it also becomes important to understand how householders perceive the waste management services provided to them, and also to better understand the perception they have about their own actual and potential roles in solid waste management.

Against this background, the present chapter will present and discuss the empirical results from our study on how households receive solid waste management services from formal and informal stakeholders, on the relationships which exist between households and other stakeholders, on the municipal assistance and support for waste management service delivery to households and on the ways in which householders perceive and evaluate or assess their relationships with other stakeholders. The specific research questions we seek to answer in this chapter are:

1. Who are the key formal and informal stakeholders in the waste chain for domestic solid waste management?
2. What kind of formal and informal relationships exist between householders and other stakeholders in the SWM chain?
3. What forms of institutional support do exist for the relationship between householders and other stakeholders in the SWM chain?
4. How do householders perceive and evaluate the actual waste services as resulting from the formal and informal relationships they uphold with other stakeholders?

In order to specify these research questions and to put them in a framework, we will consider the relationships between households and other stakeholders (formal and informal) in solid waste management chains in more detail. In the literature these relationships are discussed by several authors in different ways. Some authors (Commonwealth, 2003; Gidman *et al.*, 1995; Mugagga, 2006; Nkya, 2004; Oteng-Ababio, 2009; URT, 2009) use the partnership approach and others (Baud *et al.*, 2001; Van de Klundert and Anschütz, 2000) use the term 'alliances' to explain the relationship between different stakeholders in SWM. According to Henry *et al.* (2006) common types of arrangements or cooperation in service delivery are those between government and private sector companies, those between communities and the private sector and those between community-based organizations, NGOs and local government (Baud and Dhanalakshmi, 2007). These authors and the types of arrangements they discuss do not consider the relationship between households and formal stakeholders and neither do they discuss the relationships of households with different kinds of informal stakeholders. In addition, Ali *et al.* (2006) studied the existing linkages between public sector institutions, NGOs and the private sector, in order to promote integration in SWM. Joseph (2006) reported that the environmental problems of cities can be addressed in large part by improving the interaction between stakeholders such as national, state and local governments, the public sector, NGOs, the private sector and funding agencies. They all have a role to play to support priority actions. Joseph does acknowledge the important roles of households in waste management activities without giving a detailed analysis of the interaction of households with other stakeholders when performing waste management duties. Van de Klundert and Lardinois (1995) studied community formal and informal sector involvement in Municipal Solid Waste Management. Their study acknowledged that the two sectors tend to operate in a symbiotic relationship; however, they looked at this relationship from the perspective of municipal waste management without addressing in some detail household waste management practices.

Finally, the study of Van Horen (2004), emphasized that the relationship between formal and informal elements of urban governance systems should not be understated, given that in cities of developing countries, an average of between one-third and two-thirds of city populations live in informal circumstances.

It is noteworthy to conclude from this short review of the literature that across these range of perspectives on solid waste management, the focus tends to be on the relationships between actors and organizations at the municipal and/or national levels, while paying no or only limited attention to the role of households and householders at the local level. Building upon and adding to this literature, in this chapter we look at the formal and informal relationship between households and service providers (formal and informal stakeholders) in solid waste management chain.

Figure 7.1, presents the key elements and concepts to describe the relationships which are relevant in the context of this study. This conceptual framework provided a basis for studying the existing situation of the relationship between households and other stakeholders in waste management services.

The findings presented in this chapter address all the relationships displayed in Figure 7.1. The chapter is structured as follows: Section 7.2 outlines the data collection methods that were employed to obtain the present results. Sections 7.3, 7.4, 7.5 and 7.6 present the findings for the respective sub-questions, while section 7.7 gives the general conclusions and a discussion of the findings.

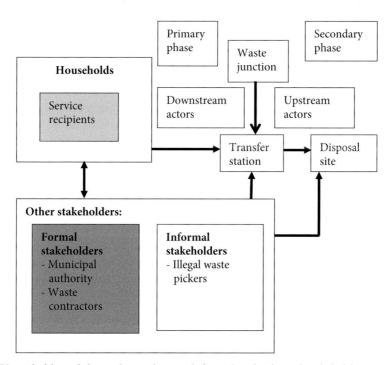

Figure 7.1. Households and their relationships with formal and informal stakeholders in solid waste management chains.

7.2 Data collection methods

This section describes briefly the methods employed in data collection as directly relevant for the topic of this chapter[27]. To answer the research questions a combination of both primary and secondary data collection methods were employed. The primary data collection used both qualitative and quantitative techniques to achieve triangulation. The primary data was collected through semi-structured interviews with key informants, questionnaire survey, focus group discussions and personal observation.

Interviews were conducted with, Kinondoni Municipal Council health officer and the Director of Tanzania National Environment management Council (NEMC) in 2008 in Kinondoni. Interviews with Kinondoni municipal health officers provided information on the formal stakeholders in waste provisioning and the role of each stakeholder, the municipal's institutional support to the formal and informal waste providers in areas under study, the municipal waste management organization. The information obtained provided answers to question (1) to (3) of this chapter. In addition, interview from Kinondoni Municipal Health Officer provided information on the current situation of solid waste management, the municipal assistance to household waste management, as well as views on how to improve households' waste management. An interview carried out with the director of the National Environment Management Council (NEMC) principally aimed to acquire information on the role of the authority in the overall solid waste management. This information enriched the information from municipal health officer.

Interviews with solid waste management service providers (CBOs and private companies) in areas selected for research aimed to generate information on their roles regarding SWM, the support they get from the municipal authority when providing services to households, how do they interact with households in service provisioning. In addition, their views on how to improve service provisioning were obtained. The interviewed waste contractors included: KWODET from Hananasifu sub-ward, CLN from Makangira sub-ward and TTM from Kilimahewa sub-ward. The information obtained provided answers to question (1), (2) and (3) of this chapter.

Through focus group discussions, households' perceptions were examined to obtain views on services provisioning by formal and informal waste providers, and to give opinions on their own roles in solid waste management in order to identify the future and potential options in solid waste management. Most of the information on illegal waste pickers was through direct observation and informal discussion with households. Informal discussion was conducted along with personal observation. The recording was done while observing and soon after the observation detailed field notes were made.

The detail of questionnaire survey has already described in Chapter 4. This chapter utilizes part (V) of the questionnaire survey on the perception of households towards waste management to answer the research question on perceptions and assessments of householders about their relationships with other stakeholders.

Various policy documents were used as secondary source of data. After obtaining primary data through interviews, various policy documents were reviewed to supplement the information from interviews. Additional information on secondary material was obtained through intensive

---

[27] See Chapter 4 for methodological details.

reviews of literature and various publications related to solid waste management. Data collected from interviews, and focus group discussion were processed and edited. These data were recorded by taking notes and transcribed before being analysed using content analysis.

7.3 Formal and informal stakeholders and their roles in domestic solid waste management

This study affirmed that both formal and informal stakeholders are crucially involved in solid waste management in the sub-wards of Dar es Salaam under study. The formal stakeholders include the Municipality, and waste contractors with legal contracts of providing collection and disposal services to households. While the informal stakeholders include unregistered and unregulated waste pickers who are engaged in collection and disposal of solid wastes from households. Both groups of stakeholders have been studied in detail and are discussed in the following subsections. The discussions will start with formal stakeholders, followed by informal stakeholders.

7.3.1 The role of municipal authorities in domestic solid waste management

The findings reported in this section are based on data gathered from observations and interviews with the Kinondoni Municipal health officer and municipal records. According to the Kinondoni Municipal health officer, the department of solid waste management is charged with operational activities of managing solid waste. The most important role of the local authorities has been their responsibility for the enactment of local by-laws and for the delivery of the service directly. The department is responsible for all activities in solid waste collection. These include planning, control, supervisory and monitoring roles. He reported that currently most of the coordination and operational programs for waste management are carried out at the municipal level because there is no sufficient supervision capacity at lower levels. The waste management sector depends on the health administration to assign health officers to work on the waste management issues at ward and sub-ward levels, where households are located. When asked specifically about the waste management tasks at the household levels, he admitted that household waste management is one of the major environmental problems in the municipality especially in informal settlements. As he said:

> Waste management in informal settlements is a problem to the municipality due to high population and poor accessibility. Most roads are not passable especially during rain seasons, infrastructures are poor. One house can accommodate six or more households (Interview, Kinondoni Municipal Health Officer, Mr Kizito, 06/11/2007).

From our discussion, he also mentioned poor enforcement of the existing by-laws as a cause of low waste collection and disposal. He confirmed that: 'To-date there is no policy for solid waste management per se'. He revealed that the problem of managing solid waste at household level is also due to an increase of petty trading. He put is as:

Petty trade generates substantial quantities of solid waste like pieces of wooden crates, plastics and papers. Many households have informal businesses indoor which increases solid waste generation from households. Some people are not satisfied by the service, hence they do not use solid waste contractors, and instead they use illegal waste pickers who in most cases do not care about the cleanliness of the environment (Interview, Kinondoni Municipal Health Officer, Mr Kizito, 06/11/2007).

On the question of waste illegal waste pickers, he insisted that:

Waste pickers are illegal, who create many problems for the municipal authority. They are the source of many problems, they scatter waste everywhere and are more nuisances (Interview, Kinondoni Municipal Health Officer, Mr Kizito, 06/11/2007).

This shows that the attitude of the municipal officials to illegal waste pickers is often very negative, regarding their role as problematic, related with dirt and nuisance, and generally not acceptable in the context of the formal, municipal solid waste management system. On the other hand, the municipal health officer acknowledges that in solving the problems of waste management at household level the municipality has appointed health officers working in the rank of waste management officers to deal with waste management issues at household level. Working together with health officers, the municipality established Ward/Sub-ward waste management plans. Sub-ward leaders are known by households as 'Wenyeviti wa Mitaa', they are the residents of the sub-wards they represent; they are elected for a period of 5 years by other residents to perform certain community management duties and to represent them in a ward (see section 7.4.1 for their duties). There are environmental committees at the ward level, which constitute the councillors at the ward and sub-ward levels. The committees make follow up on waste management activities and other environmental issues.

The municipality also collects solid waste from areas which are not serviced by waste contractors. There are designed collection points in places where municipal trucks can get access to remove the waste and transport it to the disposal site. The municipality through health officers creates community awareness and coordination and standardization of solid waste management rates. The municipality conducts awareness raising seminars to peoples on the proper waste management and payment of waste collection fees. The issues like mobilization and creation of awareness among the householders, resolving conflicts between the private organizations and the householders in relation to service performance remain the responsibility of the municipal council.

The Kinondoni health officer concluded that in household waste management, the municipality does not directly work with households, but collaborates with ward, sub-ward and waste contractors in household waste management.

7.3.2 The role of waste contactors in domestic solid waste management

The Swahili word for waste contractors is *wakandarasi wa taka*, which is used by households in the areas under study as well. According to Dar es Salaam City Council (DCC), solid waste contractors include registered private sector actors in SWM and Community Based Organizations (CBOs). In discussing the roles of waste contractors in household waste management, the interviews affirmed that, waste contractors operate under contracts given to them by the municipality, and they are expected to work according to the municipal by-laws. According to Kinondoni Municipal by-laws of 2000, 2001, and 2003, to collect Refuse Collection Charges (RCC), waste contractors are required to promote more efficient wastes collections services to their respective areas as directed by the Municipal by-laws and as per contracts. They are required to deliver satisfactory services, to collect monthly Refuse Collection Charges (RCC) from residents, to send to the local authority monthly reports on both revenue collection and performance in waste collection including reportage on the problems faced, and finally to pay refuse disposal charges.

During the interview with the KIWODET spokes lady, she said that:

> *Though Municipal Authority assigns us to collect and dispose waste, we also carry out extra duties with households in our areas of operation to raise awareness of households towards environmental cleanliness and make a good relationship so that household can pay us (Interview, KIWODET spokes lady Mrs Msosa, 8/10/2007).*

Further, she explained that KIWODET carries out thorough cleaning which include trimmings of trees in every 4th Saturday of the month. In our interviews with other solid waste contractors (CLN and TTM) we also found that households collaborate with them to clean their neighbourhoods. In fact, CLN and TTM waste contractors were not very specific on the issues of extra activities with households.

In addition, solid waste contractors were able to express the problems associated with household waste management. Common problems were found to persist in the studied sub-wards. The main problems mentioned by all interviewed waste contractors were: low payment, inadequate collections due to poor accessibility and congestion, several households in one house, presence of faecal matter in solid waste and presence of illegal pickers in the areas which they (waste contractors) operate. The concerns of waste contractors for the presence of illegal waste pickers was explained by waste contractors as:

> *Households give their waste to illegal waste pickers who often go from household to household, collecting solid waste materials from householders, and then transport to a transfer station or dump illegally. Our customers pay waste pickers for collecting their waste, and they fail to pay for our service (Interview held with Mr Komba, CLN chairman,19/12/2007).*

The above expression implies that the presence of illegal waste pickers is perceived by waste contractors as interference of their job which limits the income of waste contractors from waste. On the other hand, insufficient collection provided to households by solid waste contractors allow solid wastes to be readily available for illegal waste pickers.

Nevertheless, for all the waste contractors, low payment from households was the major issue. Not only they mentioned it as a major problem, but they also suggested how the municipality can help to tackle the problem. According to TTM chairman from Kilimahewa sub-ward, the municipality should encourage payment from households, as he said:

> *Our job of waste management becomes very difficult sometimes due to stubborn people who do not pay us. The municipality should be strong in waste management just the same as it is in collecting taxes from other businesses (Interviews with TTM chairman, 18/12/2007).*

Additionally, he suggested that the municipality should increase the number of moveable transfer stations in the sub-wards in order to allow households to deposit their waste. When undertaking this research, TTM had one truck which was used as a transfer station for households to bring their waste. The complaint of the chairman of this CBO was that one truck was not enough, and he was of the opinion that, if there were adequate stand-by trailers to be located in accessible areas within the sub-ward, the truck would be used to receive wastes from households living in difficult or inaccessible areas. Further, he suggested that sub-ward leaders should ensure that households are supervised to make sure they dump waste into the stand-by trailers. He also expressed that:

> *We have mobilized people, and everyone wants the waste collection services, but they don't see the necessity of paying. Other people complain they do not have money, but the same people spend too much on other things (Interviews with TTM chairman, 18/12/2007).*

As previously mentioned, the main concern of waste contractors in providing solid waste services to households were the low rate of payment by households, inadequate equipment, and existence of informal pickers in their areas of operation. As indicated by households themselves in survey and focus group discussion, the rate of low payment was due to unreliable and insufficient collection service provided by waste contractors. Waste contractors admitted that the unreliable and insufficient waste collection and disposal services were due to limited resources in terms of collection equipment such as stand-by trailers. Nonetheless, to solve the problem of low rate of payment, both the waste contractors and households agreed on the payment mode which they found to be easier for households to pay compared to what is prescribed by the municipal by-laws obliging households to pay per month. When these official regulations have to be implemented at the household level, they will not function if the actual situation of households is not taken into consideration. The overall picture of the waste contactors' complaints is that they argue that both social and technological problems – this case payment schedules and waste collection equipment – need to be tackled jointly in order to improve the situation.

Waste contractors raised complaints on informal waste pickers and their services to households possibly because they see informal pickers as a direct threat to their official job and income generating possibilities. The franchise system under which they operate in principle grant them the exclusive right and responsibility to provide waste collection and collect fees from households within their boundaries of operation.

7.3.3 The role of informal waste pickers in domestic solid waste management

From personal observations and both formal and informal discussions, it was found out that in the areas where there are formal waste providing services, illegal waste pickers also exist. The illegal pickers collect wastes from households and deliver the wastes to municipal stand-by trailers. A personal observation of the author as explained below illustrates a good example of informal pickers in operation.

> *Two young brothers who collect waste using a pushcart in Dar es Salaam city. Despite this place, being receiving solid waste collection from both private company and municipal truck providing the solid waste collection service; households use the services of these boys. These two boys transfer waste they collect to a nearby municipal stand-by trailer. Interestingly, even if one has no waste, households will give some money to these boys. Households identify them as 'Vijana wa Mkokoteni' meaning Youth of Pushcarts. However, these boys offer other services to households such as carrying luggage, cleaning the surroundings; also some households use them as messengers (Authors observation, 2008).*

It is was noted that these two boys visit the area daily around 17 hours and they do not allow other illegal waste pickers to move in to collect waste. By protecting their customers (households), they are sure about the source of income and know when to visit particular households to collect waste.

An informal discussion, during participant observation, with a lady in Hananasifu revealed that:

> *There are three of waste pickers who collect our waste using the pushcart or sometimes they use sacks, they can come together at a time or sometimes one or two come. We know that there is a waste contractor that has been licensed by the municipality to collect waste in our area and we have to pay TZS 1,000 per month. Waste contractors collect money at end of the month, but we don't see them coming for our waste. While waste pickers collect money on the spot and it depends on the amount of waste; sometimes we negotiate. The problem with waste contractors is that they do not reach us. Sometimes piles of waste stay for a long period and they begin to make very terrible smells. As you understand Dar es Salaam is very hot. We don't wants to stay close to solid waste pile because of the smell. When this happens, waste pickers come to rescue us, and we pay them immediately (Author's informal discussion with Mama Santu of Mwongozo sub-ward, 2008).*

It is important to note that while some of these illegal pickers of solid waste removal are filling the gap, there are some negative effects as well. Some of the waste collected is not sent to the trucks or other common points for waste collection. Instead of being properly handed over to actors operating in the secondary phase of the waste chain, the wastes removed from households are dumped in open spaces, and valleys.

7.4 The relationship between households and other key stakeholders in domestic solid waste management

As described above, the municipal council, waste contractors and informal waste pickers are the key actors and stakeholders in households solid waste management. These stakeholders were found mainly to provide collection and disposal services to households. From our earlier discussions, we saw that, there are formal and informal stakeholders who provide waste management services to households as Figure 7.2 illustrates.

The following subsections will discuss in some more detail the relationships between households and both the formal and informal key stakeholders in domestic solid waste management.

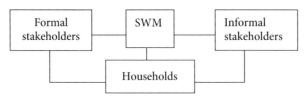

Figure 7.2. Formal and informal stakeholders' relationships with households in the context of household waste management.

7.4.1 The relationship of households with formal stakeholders

As household waste management tasks and issues are assigned to different levels of the municipal authority and some tasks are assigned to waste contractors, the set of relationships between households and other, formal stakeholders can be indicated in a way as suggested by Figure 7.3. The figure gives illustration on how the formal stakeholders are linked to one another and to households. As previously explained the municipality links up with the ward level in coordinating, supervising and ensuring follow up activities regarding solid waste management as implemented at the ward level.

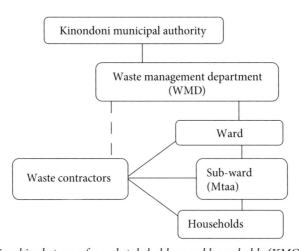

Figure 7.3. Relationships between formal stakeholders and households (KMC, 2007).

Sub-ward leaders are supervised by the ward through the Ward Development Committee (WDC) in implementing their roles on solid waste management[28]. Sub-ward leaders supervise households to implement their duties (to ensure that each household has a receptacle for waste), and mobilize households on general cleanliness. Solid waste collection zones are based on administrative boundaries, i.e. ward and sub-wards.

Duties of ward and sub-ward leaders as based on Government Notice No. 3 of 7/1/1994 include: to ensure that private solid waste collection contractors collect the waste as per agreed upon schedule, to ensure that residents in the ward pay for waste management and waste collection fees as specified in the by-laws. To ensure that private solid waste collection contractors do not charge waste management and refuse collection fees more than that set by the municipality. According to Kinondoni Municipal health officer, the waste contractors are under the responsibility of the ward/sub-ward leaders. The ward and sub-ward leaders are the ones selecting franchisee/contractors to provide waste collection service in their areas of jurisdiction.

Although *Mtaa* leaders have an official status within the formal administrative system that is specified in Government Notice No.3 of 7 January 1994 under the Local Government by-laws, they are in many ways in a hybrid position, operating on the interface between the formal and grass-root institutions (Šliužas, 2004). *Mtaa* leaders differ from the strictly formal (waste management) actors in a number of ways. Unlike the Ward Executive Officer, who is a full-time employee of the local government, often with no personal ties to the ward in question, as indicated in section 7.3.2, *Mtaa* leaders are residents of the *Mtaa* they represent, they are elected for a period of 5 years by other residents of the *Mtaa* to perform certain community management duties and to represent them at the ward level.

Roles and responsibilities of households in solid waste management are specified in Kinondoni Municipal by-law of 2001 as: to provide receptacle for waste, operate and maintain the facilities, finance the cost of receptacle for waste, clean, remove all waste and maintain cleanliness of the frontage of the house and pay for the collection fees. It is important to note that the formal relationship in household waste management exists basing on the routinely formal duties of stakeholders.

7.4.2 Informal relationships of householders with other stakeholders

In the context of this study, the informal relationships refer to the relationships or interactions which are taking place between households and illegal waste pickers when providing waste services to households. Similarly, informal relationships were found to exist between households and formal waste contractors as a results of private arrangements made between the two parties in the delivery of solid waste collection and disposal. Thus, from this study we can distinguish informal relationship into two basic forms. One is the informal relationship between households and illegal waste pickers. The second form of informal relationships is formed between households and waste contractors. These informal relationships differ from the formal relationships in one major way. Unlike the formal relationships, which result from and are regulated by the official

---

[28] The Ward Development Committee (WDC) is a ward committee which is responsible for scrutinizing and endorsing development projects before submission to the Municipal Council.

municipal waste management structures (see section 7.4.1), the informal relationships result from the negotiations and mutual agreements between households and waste contractors. The following subsections discuss in some more detail the forms that informal relationships can take.

Household and their relations with waste pickers

As explained in Chapter 5, illegal waste pickers provide waste collection and disposal services in sub-wards which are not receiving formal waste collection services, but they operate as well in some sub-wards with formal waste collection services being in place. Through personal observation during the research period it could be concluded that illegal waste pickers were mainly young men who routinely go around looking for solid wastes to pick. We observed them carrying sacks of waste on their back, and knocking on gates while asking for wastes. During the survey, some respondents indicated that they deliver waste to whoever is looking and asking for it. '*We do not have an alternative*' said a respondent from Mwongozo sub-ward during our survey. She narrated more that whenever we see '*Mkokoteni wa taka (pushcart for waste), we take out our waste, and we pay them TZS 500 per collection*'.

This means that on a monthly basis, low-income households in informal settlements pay more for the collection of their waste by illegal waste pickers, than if served by formal waste contractors against officially fixed rates. They are required to pay only TZS 1000 per month as specified in the municipal SWM by-law.

Both in the survey as well as in the focus group discussions, households pointed out that the collection service from informal pickers is more reliable and efficient than the formal service as provided by waste contractors. During personal, informal discussions one lady from Hananasifu ward claimed that:

> *These boys (informal pickers) provide security to us; provide other services such as emptying pit latrines, and acting as haulers or carriers. Sometimes I think better I give my money to these poor boys than giving it to these contractors who are rich owning vehicles, and they don't collect our waste (Informal discussion with Mama Hussein of Hananasifu sub-ward, May 2008).*

She also indicated that the majority of illegal waste pickers reside within the same or neighbouring sub-wards. She adds that, using the services provided by illegal waste-picker's is a way of supporting their livelihood, taking into consideration that the majority of illegal waste pickers depend on solid waste collection to provide for their daily basic needs. From the respondent's expressions it can be argued that, in addition to waste collection and disposal, waste pickers have personal relationships with households. The personal relationship might be established as result of being in the same neighbourhood and thus they interact frequently which enables mutual understandings, and eases communication between households and waste pickers. On their interactions, illegal waste pickers provide waste collection and disposal services to households, at the same time households support the lives of illegal waste pickers. For example, in addition to the services of solid waste collection and disposal, households use illegal pickers for other tasks which they (households) pay for as indicated in the above statement.

Households and their 'informal' relations with waste contractors

Our study found that, even though solid waste contractors have a legal contract from the municipality to provide solid waste collection and disposal services to households, they turn out to have their own agreements on the modes of operations, which are not stipulated in the municipal SWM by-law or indicated in the contract given to the solid waste contractors. Findings from the household survey and personal observations revealed that households bear responsibility for manually carrying wastes generated in their households to the transfer stations, which can take the form of stand-by trailers or solid waste trucks. This responsibility was found to be in place particularly in the settlements which are not accessible for solid waste trucks, thus making it necessary for the waste contractors to place the stand-by trailer at the passable streets within the settlement. A typical example was found in Makangira and Kilimahewa sub-wards. The interviewed waste contractors revealed that, by using local leaders, households were convinced to bring their wastes to an accessible place, where transfer stations are stationed. The TTM chairman when interviewed put forward that:

> *The majority of our customers live in valleys where wheelbarrows or pushcarts cannot reach, we had to ask these people to bring their waste along the road where we station our trucks (Interviews, TTM chairman, 18/12/2007).*

In scrutinizing the existing Kinondoni by-laws, it was learnt that the by law does not require household to transfer their wastes to a transfer station. Kinondoni Municipal by-laws of 2001 allows waste contractors to use a franchise system to collect waste fees from the households they serve. It is the municipal authority which sets the monthly waste collection fee, and households are required to pay per month. However, according to waste contractors, it has been difficult for some of the households to pay per month. During the interview with the KIWODET spokes lady it was stated that:

> *To facilitate waste fee collection, we involved households and local leaders so that we could agree on the convenient ways such as to let households pay by instalments.*

However, she revealed that some people are very stubborn to pay waste fees to the extent that they have to use health officers and local leaders to force them to pay. Some waste contractors apply the 'pay on the spot technique', whereas payment of the waste collection fee depends on the amount of waste disposed by the household. A typical example of a waste contractor using this system is CLN, a private company serving Makangira area. They use a pre-paid system called Pesa Taka system (Cash strategy meaning waste for cash per volume), whereby the households pay waste fees on the spot (per collection) when delivering waste to a waste truck. This system facilitates collection of waste fees and is both a customer and contractor friendly system. As the following quote from the CLN spokesman illustrates:

> *Our customers have accepted our services and the Taka – Pesa system is an affordable and friendly system to both of us, "customers and the company". We have enough*

support from our area leaders and customers. We have good relationship among employers, employees and our customers. We have highly motivated and committed staff. We are flexible to continuous improvement towards our services and jobs (Interviews, CLN chairman Mr Komba, 19/12/2007).

According to the CLN spokesman, usually households pay between TZS 200-300 per 10-20 kg, normally contained in one plastic bag. As we can note, in this kind of relationship, sub-ward leaders were used to facilitate the agreement between households and waste contractors. Sub-ward leaders have helped in mobilizing households in the adoption of easy methods for paying waste fees. As previous mentioned, it is the municipality which dictates the waste fee to households, also the mode of payment and impose to households through a by law without consulting households. As a result, households failed to pay waste collection fees in a way as stipulated by the municipality. Furthermore, the agreement of households and waste contractors in the delivery of waste services demonstrates that households can contribute to solving the problems in solid waste management if they would be involved in planning from the very beginning, instead of imposing requirements on them which they cannot fulfil. Households as recipients of SWM services can respond positively towards a particular shortcoming or challenge in the waste management system. When feeling excluded and not consulted however, they no longer regard themselves as having an active stake in effective solid waste management practices and therefore may decide to not take any part in certain SWM practices.

7.5 Municipal assistance to stakeholders in domestic solid waste management

Although the Kinondoni municipal council carries the overall responsibility for solid waste management as described in chapter two, the central government and national institutions play a big role and carry a considerable responsibility in the whole system of urban waste management as well. This section will describe specifically the different forms of assistance from the side of the municipal authorities for the formal actors and relationships described in section 7.4.1. The assistance which the municipality gives to solid waste management practices at household levels includes human, technological and financial resources.

In terms of technological support, the Kinondoni municipal council supports waste contractors by providing equipment such as moveable stand-by trailers. For instance, KIWODET confirmed to have received 2 standby trailers from the municipal authority in 2007, and TTM as well confirmed to have acquired 3 standby trailers in the same year. On the side of the CLN, according to the chairman, the company was offered the moveable trailers from the municipality, but the company did not accept it. '*We use our own solid waste trucks to receive waste from our customers*' as attested by CLN interviewee. However, he did not reveal any specific reasons for turning down the offer. It is important to mention that the municipal authority delivered these standby trailers to waste contractors through ward executive officers in their respective wards. It was revealed by the Kinondoni health officer that, one of the plans contained in the 2007-2012 municipal strategic plan was to supply tricycles known as Bajaj to waste contractors. These tricycles could help them to increase the efficiency in waste collection in informal settlements, since a Bajaj can easily penetrate into informal settlements due to its small size.

In terms of human resources, the municipal staff from the waste department is employed to support waste management services at the household level. As earlier discussed, the municipality assigns health officers to work on waste management issues at the ward/sub-ward levels, where the operational tasks are performed. In the course of our discussions, the Kinondoni health officer indicated that by 2008, 12 health officers had been appointed as waste management officers in 12 Kinondoni wards. The KIWODET spokes lady also affirmed that the health officer in her area of operation mobilizes households to pay solid waste collection and disposal fees, and also promotes the use of proper solid waste storage containers.

In terms of financial resources, as the Municipal health officer stated, the Municipal council pays for its waste management workers out of its general financial resources, which it derives directly from central-government grants or subsidies or from its own sources of revenues. Another most important support to households and other formal service providers from the side of the municipality is the guiding policy framework in terms of existing municipal by laws, which spells out the duties and responsibilities of each stakeholder, including that of households.

Since the municipality does not recognize the involvement of illegal waste pickers in household waste management practices, it also does not give any kind of direct assistance or support to illegal waste pickers. Table 7.1 and 7.2 presents an overview of the human and technical resources available to waste contractors. It gives the number of staff available per 2008 to the waste contractors which have been interviewed in the context of our research.

Table 7.1. Human resources available in supporting household solid waste management.

Category	KIWODET	CLN[1]	TTM[1]
Collectors using pushcarts	3	-	7
Waste fee collectors	3	5	4
Office staff	-	-	5
Waste loaders	5	3	6

[1] CLN = CLN Electrical and General Contractors Ltd.; TTM = Tua Taka Makurumla.

Table 7.2. Technological resources available in supporting household solid waste management.

Category	KIWODET	CLN[1]	TTM[1]
Pushcarts	5	-	4
Wheelbarrows	5	-	-
Waste trucks	1	3	1
Standby trailers	2	-	3
Tractor	1	2	-

[1] CLN = CLN Electrical and General Contractors Ltd.; TTM = Tua Taka Makurumla.

The details presented in the table above concern only labourers employed to deal specifically with household waste only, because waste contractors provide services to other sectors as commercial and different institutions within their areas of operation. In terms of technological resources, the situation is as shown in Table 7.2. It will be recalled that some of the equipment was obtained from municipal council.

It is worth noting that the level of equipment displayed was not sufficient in the sense that the amount of wastes to be collected surpass the capacity of the available equipment. For instance KIWODET reported that they collect 30-35 tons of waste daily. They reported to own only one waste tipper truck which has a capacity of carrying 7 tons per trip; and to the maximum it can make only 3 trips per day to the disposal site. In a similar vein, TTM reported to collect 20 tons of wastes per sub-ward per day, and they are able to transport 4 trips per week. Consequently, solid waste contractors have to hire trucks from private owners or to lend municipal waste trucks. They complained that the trucks rented out to them are in bad shape, sometimes they have to spend money on minor repairs. According to KIWODET and TTM, cost for hiring equipment ranges between TZS 40,000 to 50,000 per trip.

We can conclude by saying that the municipality gives assistance to household waste management in order to fulfil its day to day roles and responsibilities in solid waste management. Nevertheless, it is a fact that, the waste contractors have grossly inadequate equipment and too little money to purchase enough equipment, despite the support provided by the municipality. As a result, waste contractors are unable to provide sufficient and reliable collection services to large segments of the population in informal settlements.

7.6 Householders' perceptions and assessment of the relationships with formal and informal waste management actors

The perception of the households towards the present collection system was captured through the questionnaire survey and with the help from focus group discussions. The first sub-section presents the results obtained from the survey, while the second section presents the findings from the focus group discussion.

7.6.1 Perceptions and assessments as obtained from the questionnaire survey

In the survey, we examined the perceptions of the respondents regarding the existing formal and informal relationships with other stakeholders in the waste management chain. The perception of the respondents were obtained on the set of variables as indicated in Table 7.3. The respondents were instructed to select an answer that best describes their choice. So, in this case one exclusive answer was required from the respondents for each question. Table 7.3 also shows the percentages of the positive answers of the respondents on the items at stake.

When asked 'Do you need a transfer station in your neighbourhood?', 45.3% indicated that it was important to have one in the neighbourhood. This percentage is fairly high. From our observation, we realized that only 3 sub-wards were having a transfer station, while in the 3 others there wasn't any transfer station. 35.6% stated that transfer stations were not welcome since they produce an unpleasant smell; 6.1% indicated there is no need of it, and 13.1% could not give any

Table 7.3. Results of the respondent's perception regarding solid waste management (n=360).

Variable	Number of respondents	Percentage of total respondents
A need of transfer station		
It is important to have one in the neighbourhood	163	45.3%
They produce unpleasant smell	128	35.6%
There is no need of it	22	6.1%
Don't know	47	13.1%
The official responsibility of solid waste collection service		
Municipality	190	52.8%
SW contractors	125	34.7%
Informal picker	20	5.6%
Don't know	25	6.9%
Most severe problem relating to SWM in respondents household		
Public health risk	210	58.3%
Bad odour	124	34.4%
Nothing is wrong	26	7.2%
Suggested time for waste collection by respondents		
Between 6:00 and 9:00	61	16.9%
Between 9:00 and 12:00	210	58.3%
Between 13:00 and 18:00	73	20.3%
Any time	16	4.4%
Self (household) role in SWM		
Paying waste fees	290	80.6%
Bring waste to communal facility	47	13.1%
Separate waste	2	0.6%
Not willing to do any of these things	21	5.8%

answer. The question on transfer station was asked because transfer station is very crucial facility in household waste management as it integrates households (downstream actors) and secondary collectors (upstream actors). Households transfer their waste to transfer station, from where it is further transported to the disposal site.

In terms of the responsibility of solid waste collection respondents were asked 'who has the official responsibility of providing solid waste collection services to your household'? This variable was included because households play a vital role in promoting Public-Private Partnerships (PPP) with both formal and informal SWM stakeholders. They show the following scores on the alternatives offered: 52.8% of the respondents agree that the municipality should be responsible for the waste collection services, followed by a score of 34.7% of the respondents in favour of waste contractors. 5.6% preferred informal pickers and 6.9% of the respondents could not give an answer on this question. The majority of the respondents indicates that the responsibility of solid waste collection should be vested on municipal authorities. This indicates that, households still rely on the municipality to provide the waste management services, despite the fact that the municipality failed to provide the collection service for the households living in informal settlements in an appropriate way over the recent past.

When respondents were asked 'What is the most severe problem relating to solid wastes at the household level'? Poor health turns out to be the major concern of the respondents: 58.3% indicate that poor health is the most severe problem. Bad odour is mentioned by 34.4% of the respondents as the most severe problem, while 7.2% states that there is nothing wrong. Not surprisingly, thus, respondents are dissatisfied with waste collection services, and they turn out to be aware of the health risks associated with the waste problem. It was important to include this question because there is public health implication of household waste management. Also SWM has a vital role to play in the achievement of the Millennium Development Goals (MDGs), in particular. The protection of public health by means of adequate refuse collection services and hygiene is essential, especially among the poor.

Next, the respondents were asked 'What time would you like your waste to be collected from your household?' This variable was included because as we noted households are provided with door to door collection which necessitates them to bring their waste at the door step when the collector arrives. It is important then to know the time households prefer their waste to be collected because with this method of collection since waste collectors enters the premises of the household or wait at the door for a household member to bring waste out, if one is not present, the waste is not collected. The majority (58.3%) suggested the collection to be planned between 9.00-12.00, and 16.9% prefer between 6.00-9.00, and 20.4% prefer between 13.00-18.00, while 4.4% answered that any time would be okay for them.

Finally, respondents were asked 'What role would you like to play in SWM'? (they were instructed to mention the most crucial role). 80.6%, indicated that the most important role for them is paying the waste fees for waste collector to take away their waste. The possible reason for this could be that since the majority of households are receiving waste collection and disposal services from informal pickers whom households perceive to be more efficient than legal waste contractors in removing waste from households premises. As noted in the previous chapter, households pay collection and disposal fees to informal pickers on the spot, and afterwards, households don't bear any other responsibility, this shows that is an easy option to them so long they want their premises to be clean and free from solid waste accumulations. Possibly they take it as a best model.

The relatively low percentage of 13.1% who indicated that the most important role is to bring their waste to transfer the station, can be explained from the fact as we noted from our survey and observation households have negative on transfer station. The negligible percentage (0.6%) of respondents indicated that separating waste is most crucial role. We can consider that households are not aware of resource recovery and its environmental benefits. In addition the municipality has no any program for resource recovery so it is not known to the majority of households, and 5.8% indicated they are not willing to do anything in SWM, possibly due to ignorance.

The focus group discussion helped us to develop a deeper understanding and insight into households perceptions in terms of qualitative analysis. The findings of focus group discussion are as presented in next section.

7.6.2 Perceptions and assessments as obtained from the focus group discussion

Focus group discussion extends and refine the quantitative findings and illustrate with quotes from the participants. It was held with 12 participants on 8/1/2009 in Dar es Salaam[29]. The following sub-sections present the perceptions of households as obtained from focus group discussion. We asked the focus group participants to express their opinions about different aspects, actors, and relationships that were relevant for the domestic solid waste management system as they know it from their own, local experience. The findings will be reported under the categories of 'transfer-stations', 'service provisioning', 'informal pickers' and 'own role of households' respectively.

Opinions of households on transfer stations

Regarding transfer station the participants expressed different opinions, since the focus group meeting was composed of participants from sub-wards with legal waste contractors as well as from sub-wards with no legal waste contractors. As described in Chapter 5, the sub-wards with no formal collection services totally depend on other means such as burying, burning or illegal waste pickers to collect and dispose of their wastes. Five respondents from sub-wards with no formal waste contractors argued that the municipality should provide each sub-ward with a moveable transfer station, so that people can transfer their wastes to it, while one participant from these sub-wards had a different opinion as he said 'No transfer station'. In his opinion he claimed that transfer stations cause more problems. On the other hand, to show that households rely on the municipality to solve the problem he said:

> We want municipality to assign a waste contractor to our areas, and should make sure that waste contractors are paid by households accordingly (Male FGD participant, 8/01/2009, Dar es Salaam).

For those who opted for a transfer station, when asked about the location of the transfer station, they said it can be placed in any open space available, but away from their houses, and the municipality should be responsible for having it emptied on a regular basis. The participants complained about the offensive smell caused by scattered waste in their sub-wards. As one woman participant indicated:

> We are scared for the diseases which may result from waste scattered in our area. It is dangerous for our children (Female FGD participant, 8/1/2009, Dar es Salaam).

This statement shows that households are very much concerned about their and their children's health. It also shows that households oppose the placement of transfer stations in proximity of their houses for reasons of negative externalities such as health hazards bad odour and security reasons. Health hazards resulting from scattered waste are the major concern of these households.

---

[29] Refer to the methodology Chapter 4.2.3 and Appendix 6 for more details on the methodology and format used.

Moreover, a respondent from the Hananasifu sub-ward expressed his experience on municipal transfer station which was overflowing with waste and the municipal trucks were not picking the wastes, in the end leaving these areas in deplorable conditions. Supported by other respondents, he mentioned that the transfer station had also become a security threat, thugs were using them as places to hide and steal people's property and cause havoc to people. As one respondent said:

> *It was difficult to control anyone from throwing the waste in; even dead animals were dumped in, and thieves were hiding inside especially at night (Male FGD Hananasifu sub-ward participant, 8/1/2009, Dar es Salaam).*

Partly the findings here imply that availability and accessibility of transfer stations in neighbourhoods may not improve the waste management unless there is a regime in place for the proper use and the regular emptying of the waste-stations. It would also help to change the attitudes of households. On the other hand, if transfer stations are not maintained, and not emptied properly, its proximity to households becomes a nuisance or even a health hazard. Availability of well functioning transfer stations seems to be the main issue that bothers the participants of the focus group.

Householders' assessment of waste collection services

From the focus group discussion it became clear that a major concern for the participants related to the inadequate collection of wastes and the non-existence of formal waste collection services. The majority of households is not provided with waste collection services in an adequate way, and some participants do not receive the service at all. Irregularity of collection and delays, wastes remaining uncollected for a long time. Problems of waste collection from depressed lands such as valleys, were expressed by a lady from Kilimahewa sub-ward who explains:

> *There is a problem with collection vehicles reaching us living in valleys (mabondeni) areas. So we burn waste or give it to our children to dump them away from home. We have been burying and burying; now there is no more space for burying. If we bury now we dig out waste (Female FGD Kilimahewa sub-ward participant, 8/01/2009, Dar es Salaam).*

Another woman from Makoka sub-ward expressed almost a similar problem when stating that:

> *I get problems from the smoke flowing to my bedroom from neighbours waste burning practice. I once volunteered to take care of her waste but she refused and abuses me that it is none of my job (Female FGD Makoka sub-ward participant, 8/01/2009, Dar es Salaam).*

To show that they know where to forward the problems, one male participant advised the lady to take the matter to the sub-ward leader (*nenda kwa mjumbe*) to resolve the matter. On further discussions, the participants were asked whether there is any step they can take concerning the problem of the irregularity of collection services by waste contractors. To show that they know who is overall responsible, one of the respondents said:

> We report the matter to the sub-ward leader, and we stick to sub-ward leader (tunakwenda naye mguu kwa mguu) until he brings a waste contractor to come and collect our waste (Male FGD participant, 8/1/2009, Dar es Salaam).

On the issue of the preferred collection schedule, participants want waste to be collected daily if possible and at morning hours before 10.00, as one said 'usafi ni asubuhi', meaning that cleanliness is always to be carried out in the morning. By this statement there is cultural element brought into the waste practices. In Dar es Salaam and most other parts of Tanzania people sweep and clean their houses and surrounding premises before starting any other activities. Dar es Salaam city where Kinondoni municipality is located is characterized by cultural heterogeneity; it brings together large number of Tanzanian ethnic groups. People of various ethnics are concentrated in different parts of Kinondoni municipality. Many tribes in Tanzania believe that sweeping the house as a first activity of the day welcomes blessings, and sweeping when it is dark is chasing away the blessings. This brings an implication that when considering alternatives for household waste management, one must take cultural factors into account.

Householders' opinions on illegal waste pickers

On the question of illegal waste pickers, householders participating in the focus group perceive the existence of informal pickers as related to the subject of unreliable service provision from the side of the municipality. This is illustrated in the following statement by a lady who said:

> If waste collectors doesn't pass to collect our waste, we give it to illegal waste pickers (vijana wa mkokoteni) (Female FGD participant, 8/01/2009, Dar es Salaam).

Another respondent from Hananasifu sub-ward, claimed that

> We give waste to informal pickers because the waste contractor (Mkandarasi wa taka) comes to collect our waste only once per month (Female FGD participant, 8/01/2009, Dar es Salaam).

Households seem to appreciate illegal waste pickers; however, at the same time they complain that waste pickers are thieves. As said by one male respondent:

> Illegal waste pickers are important to us, but sometimes they are thieves. When they come looking for waste, they get opportunity to survey for other things. They are pickers of everything (Male FGD participant, 8/01/2009, Dar es Salaam).

Nevertheless, not all of the participants agreed that illegal waste pickers are thieves, one participant commented that thieves do not involve themselves into hardship. As he said:

> *Waste pickers are involved in carrying huge sacks of waste as well as pushing carts daily. All of these activities present obvious risks to these poor boys (Male FGD participant, 8/01/2009, Dar es Salaam).*

This quotation shows that some households are concerned with the poor working environment and the health risks of informal pickers. This demonstrates that households show a strong concern about health risks.

Householders on their own roles in solid waste management

When discussing in the focus group the perceptions about their own roles, the participants expressed the opinion that first and before all the municipality should give enough assistance with equipment such as stand-by trailers, wheelbarrows and pushcarts to waste contractors. Participants were of the opinion that they can and will pay waste collection fees if the services offered are reliable. One young man from Makangira sub-ward said:

> *We don't have a problem with paying; people resist paying for the collection of waste because municipality does not help them (Male FGD Makoka sub-ward participant, 8/01/2009, Dar es Salaam).*

However, one participant expressed his views that the service should be free of charge because some people cannot pay. As he said:

> *People are poor, how can they pay for waste when getting food for children and paying school fees is difficult? (Male FGD participant, 8/01/2009, Dar es Salaam).*

Another view was that the waste contractors responsible for the respective areas should supply proper waste storage containers to each of their customer. The container's costs shall be recovered by the contractor from the solid waste collection fees. One participant expressed his views as:

> *We should be responsible to contribute to waste storage containers (FGD participant, 8/01/2009, Dar es Salaam).*

It can be concluded that reliable waste services are a precondition for householders for them to pay waste fees. A possible interpretation of the quotes is that households in principle agree that they have a role to play in SWM services and practices, at the same time, however, our study showed that most households claims to be poor and for that reason rely on the municipal authority to offer free waste collection and disposal services. To solve problems on collection and transfer of waste, some respondents suggested that:

> *Kinondoni municipal council should allow their equipment and collection trucks to be available for leasing out in case a private contractor has a breakdown and need to borrow one, to ensure frequent collection (FGD, 8/01/2009).*

The overall picture that emerges from the focus group is that households are aware of the problems in solid waste management and want to consider solutions. Households living in informal settlements are not entirely ignored but the levels of waste service provision are considerably lower in terms of availability of solid waste management facilities such as transfer stations and frequency of collection. Householders identify the need for an institutional intervention, and they regard SWM services as primarily and principally being a municipal responsibility. In general households carry a negative perception of the role played by the formal stakeholders i.e. the municipality and waste contractors mainly because of the present inadequate removal of solid wastes.

7.7 Conclusion

This chapter confronted the research questions dealing with households as the recipients of solid waste management services. In answering the question about who are the key stakeholders involved in solid waste management, we discussed the role of legal waste contractors, the role of Kinondoni municipal authority, under its and their waste management department, and also the illegal waste pickers. An important finding in this respect concerns the role of formal versus informal waste actors. While not being officially recognized, informal waste pickers are recognized by all parties to play an important role in the collection and disposal of domestic wastes. The informal actors assume their crucial role partly in response to the absence of good quality, e.g. regular and reliable waste service delivery from the side of formal waste actors. The informal waste pickers fill the gap left open by formal waste actors. In doing so, they earn some money which enables them to meet their daily needs. Householders recognize the crucial function of the 'local' waste-pickers who tend to provide them with other services too. They support waste pickers by paying waste fees when services are being delivered. So while not being recognized and accepted as legitimate actors from the side of the official authorities, waste pickers are being recognized and used by householders on a regular basis. With respect to the formal waste actors, we found the role of ward and sub-wards leaders to be important since they are the ones supervising household waste management on behalf of the municipality and operate closest to the householders.

With respect to the question on the institutional support for waste services, we found that formal relationships in waste management are supported institutionally by the municipality in different ways. They provide technological and human resources to waste contractors, and provide the legal framework for waste services and responsibilities. It was noted in our study that the municipal authorities do not give direct technological and institutional support to households in performing their waste management practices. For householders, the municipal authorities are responsible for but not direct visible in waste local practices of waste management, despite the fact that ward- and sub-ward leaders play an important role in supervising the activities of the waste-contractors.

The question on the perceptions and assessment of householders with respect to the roles of different actors in domestic waste management, both the findings from the survey (n=360) and the focus group discussions (n=12) participants point in the same direction. Households are in general aware of and concerned about the health risks and environmental problems caused by the present inadequate solid waste collection and disposal services. They want their living environment to remain clean, and they hold the municipal authorities in principle and primarily responsible for providing adequate waste services. They bring forth a number of concrete complaints and suggestions for improvement with respect to the present, inadequate role performance of especially formal waste actors in the primary phase of the waste chains. The existing policies do not consider the local circumstances and the concerns of householders in different respects (time of collection, way of paying fees, provision of containers, prevention of nuisance and bad smells, etc.) and for that reason are assessed in a rather negative way by the householders taking part in this research.

As for their own roles in solid waste management, the householders seem to express their willingness to contribute under the condition that also formal actors take their fair share and responsibilities while recognizing and considering the demands and concerns of householders. Something of the classical chicken-and-egg dilemma seems to be at work also in Dar es Salaam domestic waste handling practices and policies.

Chapter 8.
Households and domestic waste management in comparative perspective: some findings from Kenya/Nairobi and Uganda/Kampala

8.1 Introduction

This chapter presents the results from the comparative study on household waste management in informal urban settlements in Nairobi (Kenya) and Kampala (Uganda). This comparative study was undertaken to be better able to put the results from the research in Dar es Salaam in wider perspective and to assess their validity. By comparing existing practices in these cities with our findings in Dar es Salaam, it may also be possible to identify further options for improving solid waste management. The cities of Nairobi and Kampala were selected because of their clear similarities with Dar es Salaam; they are all capital cities located in East Africa with a comparable history and holding large informal settlements. At the same time, some differences cannot be ignored such as the different socio-political, cultural and economic situation.

The comparative methodology as applied in this study did not give similar weight to the three countries/cities involved. The emphasis remained on Dar es Salaam, where most of the available time and resources were spent. The methodologies applied in Nairobi and Kampala have therefore been less elaborate, so not all the research questions studied for the case of Dar es Salaam could also be studied here. Nevertheless this comparative perspective still makes it possible (1) to put the Dar es Salaam findings in a broader perspective, (2) to identify a selected number of factors and variable which seem to be generalizable to the broader East-African context, and (3) to point out best practices in domestic solid waste management including the different roles for the various SWM-actors.

This chapter builds on fieldwork done in Nairobi and Kampala from January to June 2009. Before we elaborate on the methods applied in Nairobi and Kampala, we briefly highlight some basic characteristics of the three cities in section 8.2. In section 8.3, the methods of study relevant to this chapter are explained. Section 8.4 the solid waste management organization for Nairobi and Kampala are described in a descriptive way. The two sections to follow present the main study findings in Nairobi and Kampala. Section 8.7 compares Dar es Salaam, Nairobi and Kampala on a selected number of SWM-aspects, while section 8.8 presents the lessons learnt from comparative analysis. We conclude in section 8.9 with a discussion on the summary of the empirical findings of the comparative perspectives.

8.2 Basic characteristics of the three cities

These cities share a number of similarities: colonized by Great Britain, they are all large and fast growing cities in Sub-Saharan region, suffer solid waste management problems, and 50-70% of the population live in informal settlements. Table 8.1 shows the basic characteristics of the 3 cities.

Table 8.1. Basic characteristics of Dar es Salaam, Nairobi and Kampala.

City	Population (as per national census)	Growth rate	Area (km²)	Percentage of population in slums	Population density (persons/km²)
Dar es Salaam	2,487,288 (2002)	4.3%	1,800	60-80%	1,793 (2002)
Nairobi	2,750,561(2005)	2.8%	696.1	60%	4,230 (2005)
Kampala	1,189,100 (2002)	4.1%	195	57%	7,378 (2002)

As indicated in the table, informal settlements of East African capital cities account for over 55% of the city's population and improving SWM at household level will assist in achieving orderly cities waste management. Letema *et al.* (2010) noted that, about 70% of the East African city population lives in informal settlements beyond the reach of modern sanitation systems. This is the case in solid waste management too. What is typical to East African capital cities is that they are densely populated. Table 8.1 show the population density per person per square kilometres. This means that people in these cities live highly congested. The population has been growing rapidly since year 1900 through a process of rapid urbanization. Rapid urbanization has its implications to solid waste management sector.

As observed in Table 8.1, in Dar es Salaam, population densities reach 1,793 persons/km². It is estimated that about 60-80% of Dar es Salaam's population live in poor, unplanned settlements. Based on the 2002 Population and Housing Census, Dar es Salaam had 2,487,288 inhabitants with a growth rate of 4.3%. The relatively high population growth rate is due to increased birth rates, immigration rates, and more significantly by transient population.

According to the Kenya Bureau of Statistics, Nairobi is the capital city of Kenya with a population of about 2.7 million inhabitants with a growth rate of 2.8%, and occupies an area of 696.1 km². In year 2002, it was estimated that, some 60% of Nairobi's inhabitants were living in slums of informal settlements. Like other East African capital cities, Nairobi continues to experience influx of population from the rural areas, mushrooming of informal settlements and inadequate infrastructure for solid waste management.

Kampala is both the administrative and commercial capital city of Uganda situated on about 24 low hills that are surrounded by wetland valleys, characterized by an imprint of scattered unplanned settlements. According to UN-HABITAT (2007), the population of Kampala City is growing at annual average rate of 4.1%. This growth is influenced by migration and not just the natural rate of increase. Over 60% of Kampala's population lives in slums.

8.3 Research methodology

This section explains in brief the data collection methods employed in Nairobi and Kampala. The details of the sampling procedure and methodological approach are given in Chapter 4. Qualitative data collection methods were used in the study. The methods include semi-structured interviews with key informants to acquire information about solid waste management in informal settlements in the two cities. Semi-structured interview allowed to steer the discussion towards needed direction and ask follow-up questions in a flexible manner. The themes of the interview as well as some of the questions were predefined, some questions were formulated during the interview process.

In addition, focus group discussions were performed with selected households to obtain views and opinions regarding our main study objective. The findings from interviews were supplemented with document reviews. The sampling procedure adopted in Nairobi and Kampala was purposely sampling. The research was conducted between January 2009 and June 2009. In addition, the researcher together with experienced moderators conducted focus group discussion (FGDs).

In Nairobi key informants included the acting director from the Environmental department in Nairobi City Council (NCC), an official from the solid waste management division, an official from NEMA and selected solid waste contractors. An official from NEMA was selected because NEMA is the principal instrument of government in the implementation of all policies relating to the environment, while an official from Environmental department was selected because the Department of environment has an overall responsibility in solid waste management. The aim of interviews with the acting director from the Environmental department and NEMA was to obtain an overview of solid waste management in the city, and the service provision to households in slum areas. The interview with an official from Environmental department was carried out because the Department of environment has an overall responsibility in solid waste management. Solid waste contractors for interviews in Nairobi included the selected CBOs which are registered by NCC and provide service in slum areas Mbotela, Dagoretti, Muthurwa and Maringo. They are all low-income settlements located in informal settlements. The aim of the interviews with waste contractors in Nairobi was to obtain information regarding waste management services to households they serve. The information included the roles of households, mode of service provision and the problems encountered.

In Kampala key informants included Kampala City Council Officials, NEMA and an Engineer responsible for SWM in Kawempe division[30]. Kawempe division was purposely selected because Kawempe is the most densely populated suburb of Kampala compared to other divisions and is greatly affected by poor solid waste management. The interviews were also carried out with private solid waste contractors. The private solid waste contractors were purposely sampled from the list of Waste Collection and Transportation Firms Operating in Kampala which was provided by a KCC Engineer. The private solid waste contractors interviewed in Kampala included NOREMA Services Ltd and Hilltop Enterprises Ltd. Two active NGOs in recycling activities based in Kampala were also interviewed, i.e. Plastic Recycling Industry and Envirocare initiative. The aim of the interviews with waste contractors in Kampala was to obtain information regarding households' waste management services to households they serve. The information included the roles of households, mode of service provision and the problems they face when providing solid waste management services.

The focus group participants in Nairobi and Kampala were drawn from areas served by CBOs/private companies involved in interviews. The number of participants in Nairobi were 12, and

---

[30] The city of Kampala is divided into five administrative divisions: Central, Kawempe, Makindye, Nakawa and Rubaga.

14 in Kampala. We adopted the Dar es Salaam FGD's format (see Appendix 6). The focus group discussion was conducted in June 2009 (see details in Chapter 4). The objective of the focus group session in this study was to create a better understanding on current household SWM from the comparative perspectives of the households. Their perceptions and opinions regarding existing solid waste management were examined to obtain views on services provisioning by waste providers, and to give opinions on their own roles in solid waste management in order to identify the future and potential options in solid waste management.

8.4 Solid waste management in Nairobi and Kampala

This section describes the general solid waste management organization in Nairobi Kenya and Kampala, Uganda. The Dar es Salaam waste management organization is explained in Chapter 2.

8.4.1 The solid waste management organization in Kenya/Nairobi

According to the interviews carried out with acting director of Department of Environment, the NCC delivers SWM services through the Department of Environment (DOE) which is one of its administrative departments. The DOE is solely responsible for the delivery of these services through its Cleansing Section. As a department of NCC, the operations of DOE are regulated by the Local Government Act and its by-laws. The DOE is divided into the Administration Section and two operational sections, the Cleansing Section and the Parks Section. The Department is headed by the Director of Environment who is assisted by the Deputy Director (see Appendix 7).

The main responsibilities of the Department towards solid waste management are to: Implement NCC's SWM policies formulated by the Council's Environmental Committee; maintain public cleanliness, protect public health and the environment, and keep public places aesthetically acceptable by providing services for the collection, transportation, treatment and disposal of solid waste; to regulate and monitor the activities of all generators of solid waste; to regulate and monitor private companies engaged in solid waste activities; to formulate and enforce laws and regulations relating to SWM; and to coordinate with other departments within NCC, donor agencies, NGO's and other government organizations involved in SWM.

8.4.2 The solid waste management organization in Kampala/Uganda

The local government system in Uganda is based on the district as a unit under which there are lower local government and administrative units (Local government Act 1997). This act operationalizes the country's decentralisation policy, assigns roles and responsibilities to each level in the Local Government hierarchy and details out the role of stakeholders. The local government Act Part II 1997 equates a city to a district and Part II classified municipalities and towns as lower local governments (Okot-Okumu, 2006). Environmental Management in Uganda is decentralized as indicated in Appendix 8.

The National Environment Management Authority (NEMA) is the institution mandated to coordinate, monitor and supervise environmental Management in Uganda (Environmental Act cap 15). This is a semi autonomous body that woks closely with lead agencies (expert government institutions), NGOs and the public. The NEMA guidelines provide the framework within which the operations of solid waste management strategy will apply without impacting on the environment or degenerating it (Okot-Okumu, 2006).

According to the interview with NEMA official, the function of solid waste management was originally under the Department of Engineering before being shifted to the Department of Medical Services. This was after an outbreak of a cholera epidemic in the city. Solid waste management was assumed to be a health issue rather than a cross-cutting one. This has left out important key players such as community development officers, education, public relation and the end the Engineering department itself. Collection and transportation of solid waste in the district is a responsibility of the divisions under the Medical Officer of Health supervised by the Senior Principal Assistant Town Clerk. Figure 8.1 illustrates the institutional structure of SWM in the districts in Kampala.

Where the services are privatised, the divisions are responsible for recruitment, carrying out the complementary roles such public mobilization and awareness, supervision and monitoring of the service provider as well as topping up any shortfall in service charge collection. Where services are not privatised and handled directly, a solid waste supervisor under the Department of Health is responsible for deployment of solid waste collection vehicles, personnel, tools and equipment, etc.

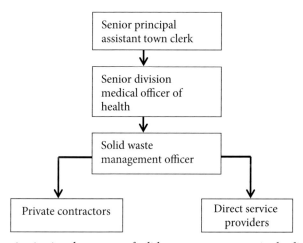

Figure 8.1. The current institutional structure of solid waste management in the district (KCC, 2009).

8.5 Empirical findings in Nairobi

8.5.1 Practices of waste handling in Nairobi: collection and transfer

The information for this part of study was obtained through face to face interview with the acting Director of Environment. Another interviews were carried out with waste contractors providing services in Mbotela estate (Kamaliza Youth Group), Riruta estate (Riruta Environmental Group) and Muthurwa estate (Muthurwa Eco Club).

The interview with the acting Director of Environment (DOE) revealed that, the Department of Environment provides most aspects of solid waste management to residents directly through its own workforce and equipment. However, in addition the council licenses private companies to provide waste collection and transportation services to households and other areas of the city. He elaborated that, NCC contract out collection services to youth groups to collect waste from slum areas and transfer waste to transfer stations owned by Nairobi City Council (NCC), and NCC is responsible to transport waste from transfer station to dumpsite. NCC requires these waste contractors to transport waste they collect from households to designated collection points or transfer station from which the NCC waste trucks pick the waste and transport it to the official dumpsite. The NCC hire collection equipment to waste contractors, at a rate of 500 Kshs. The NCC also has some informal relationships with youth groups, aimed at helping people living in slums, and promotion of composting activities and environmental clean-ups.

Households who receive waste collection services from waste contractors have to pay collection fees. However, the NCC does not state the amount of collection fees which households are required to pay. CBOs agree with their clients on waste collection tariff. According to the acting DOE, the existing by-laws do not specify the rights and obligations of the waste contractors and their customers, or specifying the standards that must be observed. The main observation made here is that the role of households as waste handlers is to provide the generated waste to service collectors.

An interview with respondent from Kamaliza youth group illustrates the waste collection practices as:

> *We provide waste bags to households for keeping recyclables waste material mainly plastics, papers and kitchen waste for the group. Kamaliza youth group operating in Mbotela Estate supply enough waste bags of 24×36cm, to households we serve, and collect every Friday. Households place bags filled with waste materials at the door step, and they pay us 100 Kshs per month (1$ = 80 Kshs) as waste collection fee to our group. Households store unsorted waste in plastic bags, plastic buckets and sometimes in boxes to facilitate collection (Interview with Kamaliza Youth Group respondent in Nairobi, 2009).*

Another interview with the spokesman for Riruta Environmental group illustrate that:

> *We started a CBO in 1999 with 300 members. We provide solid primary waste collection services in Mbotela estates a low-income area, using wheelbarrows and handcarts. Other equipment which we have include gloves, overall jackets, racks, spade and gumboots. We supply 8 waste bags per households per month to keep plastics, papers and kitchen waste. The kitchen waste we use for composting. We collect the bags with separated waste one to two times per week, usually in the second Friday and fourth Friday of the month. In turn households are supposed to pay 250 Kshs per month to the group for solid waste collection (Interview with Riruta Environmental Group respondent in Nairobi, 2009).*

The Muthurwa Eco Club respondent revealed that they collect waste from Muthurwa estate, carry out composting, farming and recycling of plastics.

Our interviews revealed that, in addition to solid waste collection services, all of the studied CBOs in Nairobi carry out recycling and composting activities out of the waste they collect from households. The important remarkable point in waste practices here is that households separate which CBOs use for composting and recycling. Further, some of these waste contractors carry out regular clean ups, educating their customers on waste management and environment management, e.g. Riruta group in Nairobi organizes a weekly clean up. Waste is collected and removed and drainage ditches are cleared with the aid of the NCC personnel and equipment. They give incentives such as Christmas cards, T-shirts to their customers.

8.5.2 Households perceptions in Nairobi

Households perceptions were obtained from focus group discussion which took place on 15/01/2009 in Nairobi. Collection and transfer of waste is one of the major concerns of households. They perceive waste management system to be characterized with irregularity in waste collection which causes the problems of waste accumulation around the containers, and attract insects and small animals. In Nairobi participants complained that the waste storage bags are vandalized by domestic animals, especially dogs and cats, which tore them in the process of looking for food, hence spreading the waste around household premises. One respondents claimed that *'informal pickers collect waste on small fees, but they dump every where at night'*. She added:

> *There is a problem with collection vehicles reaching households living in inaccessible areas. Since there are no other options for getting the solid waste collection service, these households in particularly are forced to burn, bury and dump openly their waste (Focus group discussion in Nairobi, 15/01/2009).*

It was also pointed out by another participant:

> *It is not only the city council's fault but also the attitudes and habit of the population in dealing with waste is also a problem. He mentioned people dumping their waste openly rather than putting inside the storage containers. Sometimes this is because it is the responsibility of small children to dump waste (Male respondent focus group discussion in Nairobi, 15/01/2009).*

With regard to resource recovery, households in Nairobi believed that source separation would be accepted by the householders if economic incentives will be provided to do that, such as supplying waste bags free of charge. In addition, they see the problems of resource recovery as: lack of awareness, waste collectors re-mix even if separation takes place at source, there is no advantage of sorting waste, separation is difficult and inadequate space to locate more than one container. Nairobi households suggested separation should be at CBO point of sorting or at transfer stations; the city council should build transfer station with chambers for different types of waste incentives should be given to those who separate waste, as one participant said:

Composting should be at community level. If containers are provided segregation can work. He finally raised a questions why is sorting done by few households? (Male focus group participant, 15/01/2009, Nairobi).

Regarding transfer station for Nairobi households they suggest each estate should be provided with a skip of which NCC should be responsible for maintenance and emptying. They emphasized that NCC should provide secured and fenced transfer stations. Households seemed to blame the NCC for the problems they face. As indicated by one participant:

Solid waste management is not a priority to Nairobi City Council (NCC). NCC should treat waste as other public services. Informal pickers collect waste on small fees, but they dump everywhere at night (Focus group discussion, Nairobi).

8.6 Empirical findings in Kampala

8.6.1 Households as waste handlers in Kampala: collection and transfer

The information for this part of study was obtained through face to face interview with four respondents: the Kampala City Council (KCC) official dealing with solid waste, Hilltop enterprising, Plastic Recycling Industry and Enviro Care. All the interviews were carried out in Kampala city.

The first interview was carried out with the KCC official. According to the KCC official, in Kampala, solid waste management in the district is a responsibility of the divisions under the Medical Officer of Health supervised by the senior principal assistant town clerk. The council is responsible for management of the sanitary land fill whose operations have been contracted out. Solid waste contractors operate under open competition. Interviewee acknowledged that in low income areas and slums there is problems with solid waste management because waste contractors do not want to go there to collect waste. The first problem could be due to the accessibility to slum areas where waste is.

The second interview was carried out with the Hilltop enterprising manager. On the question of solid waste collection he revealed that, households wait for the company vehicle store pour the their waste. Used plastic bags, old containers and boxes are used by households to carry waste to the vehicle. The waste fee from households is negotiated between the waste contractor and the KCC, because KCC have their own ceilings. According to the manager; in very low areas such as Katanga they charge 1000 Ugandan shillings only per household per month, but it is also difficult because people claim that they do not have money. In other areas especially middle income households they charge 10,000 Ugandan shillings. According to him, about 10% of the residents here in Kawempe division pay for the waste collection services. The division authorities are supposing to be paying the top up but sometimes they actually do not pay most times, he insisted that:

Kawempe division is supposed to give us a top up because it is sometimes difficult to collect money in poor areas, so the money from the government is supposed to subsidize the poor. The money households pay is something that we negotiate with KCC because KCC have their ceiling (Interview with Hilltop enterprising Manager, Kampala, June 2009).

Regarding the skips by KCC, the waste contractor said that:

> The skips have been removed in most parts of Kampala because they became expensive
> to maintain. Skips are good and I like them because when they are placed in different
> areas, a vehicle just comes to pick. First you do not need many people to work unlike
> other methods but also it saves on time. In the system of skips, you could take like ten
> trips a day unlike today where you can take about two trips a day. The absence of skips
> has also worsened the situation because now people dump garbage anywhere because
> they have no containers to store the garbage (Interview with Hilltop enterprising
> Manager, Kampala, June 2009).

The third interview with waste contractors in Kampala was carried out with the Plastic Recycling
Industry Manager. He revealed that they provide 60,000 households in Kampala for the purpose
of sorting plastic bottles to be brought to the company. They are doing the sensitization to make
sure people really separate the waste and especially the plastic bottles. As an incentive:

> We give out bicycles using local councils to people collecting plastics and they have
> monetary incentive for those who may work hard and collect bags of plastics to them
> (Plastic Recycling Industry Manager, Kampala, June 2009).

The fourth interview with waste contractors was carried out with Enviro Care Chairman NGO
based in Kampala. He attested that, they are engaged in sensitizing of the communities and the
physical removal of waste. They collaborate with private manufactures that sponsors them by giving
the incentives like sugar, soap which they give to people that help them in the collection of solid
waste. They give incentives to people who help them to collect waste so as to instil discipline and
help people develop a culture of cleanliness to avoid littering of waste. According to the chairman,
people have been very supportive. They provide bicycles and wheel barrows to households to use
in the removal and transportation of garbage from one place to another.

8.6.2 Household perceptions in Kampala

Households perceptions in Kampala were obtained through a focus group discussion which took
place on 19/06/2009 in Kampala. The main concern of households was on waste collection. They
suggest that, KCC should revive the skip buckets to allow people bring their waste so that it can
be collected by private waste contractors or KCC.

> KCC can revive the system of skip, provided that they should be removed daily or after
> every 2 days (Female focus group participant, 19/06/2009, Kampala).

On the contrary, they point out that the local communities have rejected KCC skips from being
placed in their areas because of the unreliable collection schedules by KCC.

With respect to resource recovery, Kampala households view the problem of resource recovery
as: lack of information to people and enough sensitization, waste is not sorted at the source,

composting at household level is not possible. Therefore, they suggested that composting should be carried out by the municipality, at parish/community level. However, they believed that source separation would be accepted by the householders if monetary incentives will be provided.

> *If you give us money we will separate the waste (Female focus group participant, 19/06/2009, Kampala).*

Moreover, Kampala households blame KCC for not being responsible for waste management.

> *People refuse to give their waste due to financial problems, so they dump it at the sides of the roads (Focus group participant, 19/06/2009, Kampala).*

Another respondent pointed out that:

> *KCC does not care much about solid waste. Skips were removed because KCC was no longer responsible. KCC can revive the system of skip, provided that should be removed daily or after every 2 days. The local communities have rejected KCC skips from being placed in their areas because of the unreliable collection schedules by KCC (Female focus group participant, 19/06/2009, Kampala).*

For this expression households blame the local council for not being responsible in solid waste management, on the other hand KCC claims people for improper management of solid waste. The households of Kampala expressed problems with careless attitudes towards solid waste.

> *In Kampala, the people do not care about waste. They discard it on the streets and these make problems that we suffer from (Male focus group participant, 19/06/2009, Kampala).*

8.7 Comparing Dar es Salaam, Nairobi and Kampala on a selected number of solid waste management aspects

The major findings in the selected number of aspects in the domains studied are compared for the three cities in terms of service providers (the main actors solid waste management), waste handling (flow of waste to transfer station) and perception of households regarding services they receive (role of household as end-users in solid waste management chain). As already discussed in methodological chapter (Chapter 4), the main method used for obtaining information were semi-structured interviews with key informants and literature review. These methods generated information on the main principal actors in solid waste management. Information for comparison on the flow of wastes to transfer station was obtained through interviews with waste contractors. The role of households as end-users in SWM chain was studied through focus group discussion.

8.7.1 Solid waste management organization

The waste management organization they all operate within the framework of decentralization. Decentralization policies assign roles and responsibilities to each level in the local government hierarchy and details out the roles of each stakeholder.

With respect to formal structure of solid waste management, Kinondoni municipality has a well established waste management department which deals with solid waste management in which the lower authorities of the municipality, i.e. ward and sub-wards are the ones supervising household waste management on behalf of the municipality and work closest to the householders. For instance as mentioned in Chapter 7, sub-ward leaders supervise households to implement their duties. For instance one of the duty of sub-ward leaders is to ensure that each household has a receptacle for waste), and mobilize households on general cleanliness. Kinondoni municipality formal solid waste structure dictates roles and responsibilities at each level of municipality, including requirements of private solid waste contractors and households.[31]

According to the interview with the acting Director of Environment, the Department of Environment in Nairobi City Council is responsible for SWM in Nairobi. It was established in 1996, since then the department has had a mandate to carry out waste management in Nairobi city. According to the acting Director the role of the department is to enforce and implement laws and by-laws. Also the department issues licenses needed to run waste-related businesses.

The department of Environment is divided into 3 divisions: Environmental Planning Management (EPM) Division, Solid Waste Management (SWM) Division, and Management of Parks and Open Spaces (MPO) Division (see Appendix 7). According to the acting Director of Environment at NCC, the main task of the EPM division is to take care of environmental legislation in cooperation with NEMA.

In specific the department carries out the following duties: implement NCC's SWM policies formulated by the Council's Environmental Committee; maintain public cleanliness, protect public health and the environment, and keep public places aesthetically acceptable by providing services for the collection, transportation, treatment and disposal of solid waste; to regulate and monitor the activities of all generators of solid waste; to regulate and monitor private companies engaged in solid waste activities; to formulate and enforce laws and regulations relating to SWM; and to coordinate with other departments within NCC, donor agencies, NGO's and other government organizations involved in SWM.

8.7.2 Service provision to households

The aspects which we compare under service provisioning include formal waste collection, practices of waste handling after it has been generated by households.

The practices of handling waste after it has been generated which we discussed include the practices of storing wastes, waste separation, and ways of (non) handing over the wastes to both formal and informal collectors who connect households to the secondary phase of the waste chain.

---

[31] See Figure 7.1.

As already mentioned, the 3 cities embarked on privatization by granting a license for waste collection and disposal to private waste contractors. In all the three cities waste collection was privatised in the 1990s. In all the cases, the privatisation was a result of the serious failure of the municipality to collect household waste. However, the modalities of privatisation are different in the three cities, especially due to the differences in local council mode of operation.

Kinondoni municipality has a franchise agreement that allows the waste contractors to collect waste fees, the Refuse Collection Charges (RCC). These charges are collected by waste contractors directly from households receiving waste collection services. This agreement relies on by-laws that the households are subjected to (see Chapter 7.4.1). The obligations of the waste contractors and households are specified in the municipal solid waste management by-laws. The franchise agreements oblige waste contractors to provide collection services even to households from poorer areas, and the ownership of the waste lie with the waste contractor, and it remains the responsibility of the municipality to supervise, coordinate and control waste contractors. Additionally, the municipality must maintain some degree of oversight and or set limits on the type and level of tariffs. Kinondoni municipal authority sets out the monthly waste collection fee, and households are required to pay per month.

Nairobi, in contrast, the local authorities opted for a management contract, which is a less demanding form of privatization. Tax collection remains the preserve of the municipality. NCC continues to directly levy taxes on users of its services, and pays the private company separately. The institutionalization of the role of waste contractors is not as clear as the Kinondoni. In comparison, the mode of Kinondoni state clear institutional roles of waste contractors, and avoid disputes with the role of the formal stakeholders. Whereas for Nairobi case, waste contractors remain unguided, operating without any institutional or legal regulations. In other words, there is no legal backing or official support to grant private waste providers exclusive rights to the solid waste collection services in informal settlements and to require households to participate. Notwithstanding, many small private sector service providers are participating informally and it has been difficult to establish their impact. In Nairobi, households living in slum and poor areas are mainly receiving waste collection services from Youth Groups. The Youth Groups engage in other services such as composting to increase the income. The waste collected is transferred to transfer stations owned by NCC, and the removal of waste from transfer station is the responsibility of NCC.

Likewise in Kampala, there is no formal engagement with the private contractors and so they are very difficult to supervise. The firms are not formally operating in the division. There are no terms and conditions against which these firms operate. No any formal relationship between the division and contractors exist. The division therefore, has no basis on which to supervise and monitor firms.

In general, in Nairobi and Kampala waste contractors are operating in open competition purely on business competitive basis. Waste contractors simply obtain a business license and start offering solid waste collection services, without vetting or regulations. This cause many households from poor areas not to access waste collection services. There are no by-laws specifying the rights and obligations of the companies and their clients, or specifying the standards that must be observed.

8.7.3 Households perspectives

Households perceptions were explored through focus group discussions conducted in Dar es Salaam, Nairobi and Kampala. Thus, it was possible to identify common views pertaining to solid waste collection and disposal services delivered to households, and also perceptions of households on their own roles in SWM.

An important general observation from households was that, although in all the 3 cities the waste collection and disposal services are provided to households by private solid waste contractors, the respondents felt that the local authorities should be responsible in solid waste management. This shows that households are not happy with the private sector. The reason for this could be because their neighbourhoods are not yet clean. The other possible reasons for this could be that, households living in informal settlements are poor, and do not have enough money that enables them to pay for waste collection and disposal services taken into considerations that they have other priorities like school fees for their children, food and healthcare. Usually, poor households' have very irregular income which do not suffice for basic needs, leave alone other requirements. The other possible reasons for households to believe that SWM is the responsibility of local authority despite the fact that the 3 local authorities failed to provide waste services to a large population in informal settlements could be that, households do not feel particularly concern with SWM issues, all they want is to have their solid waste removed from their premises. Moreover, actually, households were accustomed to free solid waste management services from the local authorities. As local authorities usually have overall responsibility in the management of solid waste, it may be normal to households to trust that for them (households) don't have any responsibility other than receiving the free service. The following quotes of households illustrates the common beliefs of households from 3 on local authorities roles:

> *Municipalities should allow their equipment and collection trucks to be available for leasing out in case a private contractor has a breakdown and need to borrow one, to ensure frequent collection (Dar es Salaam focus group).*

> *City council should build transfer station with chambers for different types of waste and remove collected waste from skip buckets (Nairobi focus group).*

> *KCC should revive skip buckets for us to dump our solid waste, KCC should remove collected waste from skip buckets. If skip bucket are provided waste should be removed daily or every after 2 days. (Kampala focus group).*

But it was interesting to find out that households in general from the three cities had pre-conditions for them to participate in SWM. The following quotes as captured from the respondents illustrates it:

> *If you give us money we will separate the waste (Kampala focus group).*

> *If containers are provided segregation can work (Nairobi focus group).*

The Dar es Salaam households believe that there is no use of separating waste at household level because the waste collectors will re-mix it.

> *To separate waste is wasting of time, because where it goes will later be mixed (Dar es Salaam focus group).*

From these quotes, we want to argue that households expects some incentives for them to play certain roles in SWM. Another important general observation as obtained through focus group discussions was the concern about transfer station. Households acknowledged that, it is important to have transfer station in their neighbourhoods, however, their highest concern on transfer station was on the health effect and pollution. As they said:

> *We prefer movable transfer stations in streets where households can deposit waste. These transfer stations should be supervised to make sure that waste is put inside the trailer (Dar es Salaam focus group).*

Kampala participants indicated that:

> *If possible households should take waste to skip buckets then, KCC transfer to dumpsite to avoid nuisance (Kampala focus group).*

And Nairobi participants stated that:

> *Each estate should be provided with skip buckets as it was before. If skip buckets are provided and not taken care by NCC even dead animals are dumped in (Nairobi focus group).*

The concern of households on transfer station illustrates the relevancy of transfer station in household waste management, as illustrated by our theory of MMA (see Chapter 3).

8.8 Lessons learned from the comparative analysis

While in section 8.3 the solid waste management aspects which were compared in Dar es Salaam, Nairobi and Kampala are discussed, in this section we will discuss lessons learnt from these cities on the basis of the findings from the study. By looking at the solid waste managements aspects from the comparative analysis, we aim to gain a better understanding of the present situation and to identify the key SWM aspects relevant for improvements in the waste management practices of households in Dar es Salaam, Nairobi and Kampala.

The overall picture that emerges from our empirical research confirms the widely recognized fact that there is a lot to be improved in the present situation. The first important lesson drawn from these cities is that households do not only generate waste, they are also involved in waste handling practices. These practices include storage, waste separation and transferring waste to transfer station. With respect to the practices of storing wastes in the three cities seem to be very similar, different type of containers were used. From my own observation, and interviews with solid waste contractors and from focus group discussion commonly used storage containers include plastic bags, old containers, boxes. This was clearly the case Dar es Salaam, Nairobi and Kampala as well.

With respect to the practice of waste separation at source, the study revealed that, the cases (Dar es Salaam, Nairobi and Kampala) have no policy or programs that encourage waste separation at household level. The study, however, showed that in these cases households practice waste separation in different manners. In Dar es Salaam some households keep separate recyclable materials such as plastic and glass containers, and do not mix them with other everyday household wastes. The Nairobi and Kampala cases showed that households separate domestic waste at source into organic matter, recyclables and general garbage (plastics, papers, glass and kitchen waste) for the waste contractors. In the Kampala case, we revealed that waste contractors use households to collect plastics for them. They provide incentives (monetary, wheelbarrows and bicycles). The important remarkable observation regarding waste separation is that in Nairobi households were given waste bags by waste contractors to contain their separated waste.

Waste separation is pre-condition for the composting process in SWM. Our field work revealed that composting activities are carried out by waste contractors in Dar es Salaam and Nairobi. In Kampala composting was not reported to take place at any level. Waste for composting in Nairobi and Dar es Salaam are obtained from households. It has been acknowledged that composting activities can significantly reduce waste volume by 50-60% (Yhdego, 1993). Taking into consideration that over 60% of waste in developing countries is organic waste; therefore we can conclude that household practices of separating and supplying waste to waste contractors are contributing to reduce the overall costs of SWM, though this is not recognized by official authorities.

With respect to the practices of collecting and transferring domestic waste, a lesson learnt in this aspect is that in all the three cities waste which is stored in the households is transferred to the transfer station.

In Dar es Salaam we learnt that there are two systems: CBO employ waste collectors to collect and transfer waste to transfer station, the second system householders are involved in bringing their waste to the transfer station. In the Nairobi case, householders are not involved in bringing waste to transfer station; waste contractors collect waste door to door from households and transfer it to a transfer station; households are not responsible for the task of transferring waste to transfer station.

An important remarkable observation is that private waste collectors in the 3 cities have opted for using technologically simple modes of solid waste collection (working with pushcarts, wheelbarrows) without basing on the use of sophisticated technologies. These methods applied in Dar es Salaam, Nairobi and Kampala are both cheap and fit local conditions (informal settlements) and have the additional advantage that do not have large capital assets. In all the cases, we explored that the private sector is involved in primary waste collection by using handcarts and wheelbarrows to collect waste from inaccessible locations through door-to-door collection, and waste trucks to transport the wastes from the transfer point to the dumpsite

Public-Private-Partnership has been promoted in the three cities between local authorities, solid waste contractors and households. Privatisation has played an important role in the household solid waste management (HSWM) as witnessed in the study cities. This comprises formal and informal stakeholders. Looking at the formal sector, Dar es Salaam, Nairobi and Kampala have initiated the privatisation of solid waste collection and disposal services under franchisees type of contract.

The privatisation of solid waste management solid waste collection services have enabled local authorities and service provision closer to households. In Dar es salaam, Nairobi and Kampala we noted that several SWM activities are moulded around partnership. The nature of these arrangements depends on the activity concerned. The two cases of PPP can be noted. The first case is the legalized PPP which is partnership between the municipality and waste contractors; while the second case of PPP is not legalized one which is between the waste contractors and households (service recipients). The first case of PPP is governed mainly by a legal contract that defines the partnership and specifies the roles and responsibilities of the partners (municipality and waste contractors) involved. The second type of PPP emerged from the joint efforts of households, waste contractors and local leaders in managing domestic wastes. In all the three cities we noted that waste contractors (waste collectors) and households have mutual agreement on the modes of operation which are not indicated in the contract given to the solid waste contractors.

It is important to note that, the partnership found between households and other stakeholders result from the consensus outside the formal waste management system. For example in Dar es Salaam, households take the responsibility of transferring waste to transfer station on the basis of their own agreement with waste contractors. They also agreed on the convenient ways of waste fee collection after failing to cope with the one indicated in the contract. This type of partnership contribute to improvement of SWM aspects at household level, by supporting existing patterns of collection and disposal services. Both in Nairobi and Kampala the notable established type of PPP is between local authorities and waste contractors. The other partnership is between waste contractors and households on agreement of waste separation in managing domestic waste noted in Nairobi and Kampala. Waste contractors have encouraged households to separate kitchen waste, paper, and plastics at the source effectively reducing the quantity of waste collected for final disposal. The waste contractors in Nairobi and Kampala who have prompt the involvement of households in waste separation have initiated the recycling projects and composting scheme (the Nairobi case). This type of partnership contribute to the improvement of waste separation at household level and thus to cleaner household premises. Further, such partnerships contribute to the reduction of waste flows for the final disposal. Studies by Baud *et al.* (2001) and Post and Baud (2004) assessed the different types of partnership, however at municipal and city levels. Our study is the first to document different types of PPP in solid waste management at household level in East Africa.

Of importance in this case is that in these partnerships the public sector is the key player. The same observation was reported by Post and Baud (2004). The local authorities usually take the prime responsibility not only in organizing, but also to monitor the performance of waste contractors. A distinguishing characteristic of partnership, however, is that they directly or indirectly serve the interest of the public (households).

8.9 Conclusion

This study has generated some data and information that would be useful for future planning and management of solid wastes by policy makers in the three cities. The comparative study carried out indicated major problems and challenges faced in waste management services for households living in informal settlements. The empirical findings revealed several factors that cause problems in solid waste management in areas under study.

Comparing the three cities with regard to solid waste management organization all are in line with the decentralization process that has been adopted by the respective countries. This has allowed the involvement of other stakeholders (private waste contractors) to provide solid waste collection and disposal services to households that were traditionally the government responsibilities.

With respect to waste handling (storage, collection and transfer), the three cities are similar in a way that the problems and constraints encountered in each case cut across. With the practice of resource recovery, none of local authorities is recognizing it, and waste separation at household level is not supported. With this observation, therefore, the identified shortcomings in these SWM aspects need to be (re)considered in the context of future policies on domestic solid waste management in informal settlements in East African capital cities.

Households can do a lot to improve solid waste management when establishing partnership with other stakeholders who can bring in essential complementary resource. A typical example of complementary resource in our case is the action of waste contractors in Nairobi of supplying waste bags to households to keep the separated waste materials. Our study on the selected SWM aspects in Dar es Salaam, Nairobi and Kampala has clearly shown that partnerships are relevant and need to be taken into consideration in academic and policy circles. Although we observed different types of Public-Private-Partnership after privatisation in the three cities, constraints to better performance related to household waste management in various SWM aspects in informal settlements remained. A major question, regarding these partnerships, therefore is what can be done to safeguard the public interest (households) within the different types of partnership? This means a future research is necessary on such partnerships to obtain a better understanding the necessary provision with regard to the different partnership arrangements which can contribute to the improvement of solid waste management services in low income areas of East African capital cities. With this observation on the role of PPP, we aim to illustrate the basic notion of the theory of MMA in particular with respect to its claim that the involvement of end-users (this case households) in solid waste management practices.

From the perceptions of households regarding waste management services provided to them we can conclude that: In all three cases (Dar es Salaam, Nairobi and Kampala), households are not entirely ignorant on waste management services, they are concerned about the health risks and environmental problems caused by the present inadequate solid waste collection and disposal services. They want their living environment to remain clean, and they thought that the local council are responsible for providing adequate waste services. They were able to raise a number of complaints in several aspects of SWM, suggestions for improvement with respect to the present, situation. Therefore, in future policies can consider how to integrate households concerns in the processes of improving waste management services. The existing policies do not consider the local circumstances and the concerns of householders in different respects. As for their own roles in solid waste management, the householders seem to express their willingness to contribute under the condition that also local authorities play their roles and responsibilities while recognizing and considering the demands and concerns of householders.

The information obtained from the three cities regarding household waste management makes it possible to identify relevant factors to be considered when organizing future interventions in SWM practices at the household level in these cities.

Chapter 9.
Conclusion and discussion

9.1 Introduction

This study has been conducted in the context of the PROVIDE programme which addresses the need for sustainable and accessible infrastructures for sanitation and solid wastes in East Africa. In this interdisciplinary project, both technical and social scientists cooperated in order to achieve an integrated perspective to sanitation and solid waste management. From the division of tasks agreed upon in the PROVIDE programme, this thesis has been assigned the role to investigate the situation of solid waste management from the perspectives of households and in particular of households living in informal settlements in Dar es Salaam city. The focus on households was chosen from the conviction that policy makers and technologists could benefit from more detailed knowledge on the waste-behaviour of households when planning new socio-technical interventions in the solid waste practices to be found in informal settlements.

Up until now, the different roles conducted by (poor) households in the Solid Waste Management Practices (SWMPs) in informal settlements in East Africa have not been confronted in the literature in any detail. Households so far have been considered primarily in terms of the passive recipients and users of waste services. In order to better understand and analyse the contribution of households in solid waste management chains, a conceptual framework has been developed which explicitly incorporates and highlights the roles of household as crucial and active elements in waste chains. The conceptual framework used in this study for the analysis of the role of households in solid waste management is inspired by and partly based upon the theory of the Modernized Mixtures Approach (MMA) which underscores the PROVIDE project. This MMA was elaborated and adapted in order to make possible the study of solid waste management practices from a households perspectives. As a result, this study contributes to the PROVIDE program and its theory of the MMA by demonstrating in great detail the contribution of households in actual solid waste management practices in low income areas in capital cities in East Africa.

The empirical chapters of this study focused on six selected sub-wards in Kinondoni municipality in Dar es Salaam city. They were selected because they are high density informal sub-wards. These sub-wards are categorized as informal settlements because they are characterized by high housing densities, an unstructured road infrastructure and inadequate provision of water, electricity, and sewerage services. However, they consist of residents with varying socio-economic backgrounds, specifically with respect to income, education, and type of employment. Even though the government of Tanzania has made initiatives to improve the situation of solid waste as explained in Chapter 2, most households in these informal settlements face many problems resulting from the inadequacies and inefficiencies of the present regime of solid waste management.

Three clusters of questions on the roles and responsibilities of households in solid waste management practices have guided this research. Firstly, how to define domestic solid wastes and their main characteristics? Here we look into the flows of household waste from a technological point of view in particular. We try to specify domestic waste flows with respect to their volumes and compositions. How much domestic waste is generated within families and per capita and what can be said about the factors behind a particular composition of the domestic waste flows. We try to determine the social factors influencing some of these waste characteristics at household level. The second set of research questions deals with the everyday life behavioural practices of handling domestic wastes. Practices of different kinds, enacted by different members of the household. So we investigate what kind of waste management practices are currently being applied by households. Who is responsible for the collection, storage and transport of domestic wastes; how do the domestic solid waste flows 'travel' from the household to the transfer station, and which other relevant strategies of householders concerning domestic solid wastes can be discerned in case they cannot get rid of their wastes in a regular way. The third set of questions refers to the position of households as stakeholders in the more encompassing chains for handling domestic solid wastes. We investigate the perception and assessment of the waste services provided to the household by non-domestic actors. Here we look into the ways in which households are being served by both formal and informal providers of waste-services. How do householders perceive of the roles and the performances of both the formal and informal stakeholders in the handling of domestic wastes. What kind of institutional support does exist for SWM practices and for developing a relationship with formal and informal stakeholders.

These three clusters of questions have been approached from a social science – and more in particular from a sociological – point of view, using the theory of the Modernized Mixtures to develop a conceptual framework for specifying the research questions. Section 9.2 first recapitulates some key elements of the conceptual approach used in the study, while the findings for the specified research questions are reported in section 9.3 to 9.6. When presenting the main findings in summarized form, we try to move beyond just reporting on the data. We try to put the results in a broader perspective and discuss the findings also from a theoretical point of view. We end with sections 9.7 on overall conclusions and implication for future research.

9.2 The Modernized Mixture Approach as a framework for assessing domestic solid waste practices in informal settlements in Dar es Salaam

The basic concept of the Modernized Mixture Approach was used to develop the conceptual framework for studying solid waste management from the perspectives of households. The framework was discussed in the introduction and presented in some detail in Chapter 3 and in Figure 3.4. Here we shortly recapitulate the key emphases of the Modernized Mixture Approach as being of direct relevance to our main research findings.

First, the MMA combines technical and social aspects into one framework. When looking at solid wastes practices at the domestic level, both technical aspects concerning the waste flows themselves as well as social aspects concerning the actors involved in the handling of the wastes are taken into account. By combining technical and social factors and dynamics, a technological reductionist view on waste handling by households is avoided while making room for more in depth understanding of the interdependencies between technologies, actors and existing infrastructures.

Second, the MMA represent a mixture of technological and social elements which is the result of trying to adjust (also solid waste) infrastructures to the local circumstances. In the light of local needs and circumstances, the infrastructures should be designed, developed and managed in such a way that conditions for access are optimal and that the infrastructures are able to withstand and adapt to changing local conditions also in the next future. For our particular empirical case of Kinondoni municipality in Dar es Salaam city the notion of access means that also the significant group of low income people must have access to solid waste services. This so called 'pro poor sanitation' argument turns out to be very relevant in our empirical research, as will be reported in the sections to follow. In fact the basic conditions in the informal settlements were conditions of insufficient provision and lack of proper management in many cases. By the application of MMA in the areas under study, we elaborated this pro-poor dimension of the MMA while aiming to contribute to the improvement of solid waste management by creating solutions which takes into account the specific local situation of 'under-servicing' major parts of the poor population in particular. The accessibility criterion was linked to the issues which relates to the performance of stakeholders in providing the service. These issues include resources in terms of financial, technical and human labour necessary in solid waste management. These factors may influence inclusion or exclusion of households from receiving the service. Also the physical environment of the area such as congestion and narrow streets may affect the accessibility by waste contractors having difficulties in entering the settlements and householders have difficulties getting easy access to waste transfer points.

Third, as discussed in the introduction of the thesis and documented by Spaargaren *et al.* (2006), the performance of a specific modernized mixture is to be judged against a set of three criteria: ecological sustainability, accessibility and flexibility. When trying to assess the performance of existing (solid waste) infrastructures at a particular place, the three criteria should in principle be applied in a combined way, looking for the best 'trade-off' between scores at the different dimensions. As will become clear from our results, the criteria of accessibility and flexibility or robustness have been dominant over the factor ecological sustainability. The ecological sustainability indicators were not developed during the field work due to research priorities against the background of time constraints. In this study, the ecological sustainability criterion encompassed mainly the issues related to the existing waste management practices (waste re-use, recycling, reduction) by households. Also the practices of open dumping, burying and burning as reported in our empirical chapters have serious implications for ecological sustainability. While ecological sustainability does play a role then in our findings, they have not been made the key criterion for the research design since being determined to a considerable extent by the other criteria of accessibility and flexibility. We will come back to this at the concluding section.

9.3 Households as waste generators

This section intends to answer the first research question which was addressed in Chapter 5. The question reads as: 'What are the characteristics of household waste'? The question was answered by looking at the: per capita generation, composition of household solid waste and factors influencing waste characteristics at household level. We examined the role of household as waste generators by looking at the waste characteristics through a waste characterization study. In this study we examined 360 households drawn with a random sampling procedure. The underlying factors that influence per capita daily waste generation and waste composition were identified and discussed as well in order see how technical, social and economic factors related to households trigger the amounts and the kinds of wastes generated by households.

9.3.1 The per capita daily waste generation

The average per capita waste generation for each studied sub-ward was presented in Chapter 6. Figure 9.1 presents the values in a bar chart.

The average per capita waste generation of the whole studied selected sample is 0.44 kg/cap/ day when not specified for socio-economic categories. When the socio-economic differences are included in the analysis, the per capita waste generation varies, in low-income sub-wards this was 0.36 and in the middle income 0.52. In the case of per capita daily waste generation by income category it is the medium income that generates the most waste. These findings are comparable with earlier studies. According to previous studies conducted in DSM city the average per capita generation rate of domestic solid waste was estimated at 0.42 kg/cap/day (Kaseva and Mbuligwe, 2005) and 0.39 kg/cap/day (Kaseva and Gupta, 1996). However, these studies did not specify whether the waste characterization was conducted at source of generation, i.e. at the household level, at the transfer station or at the landfill. Many solid waste characterization studies have been performed at transfer stations or landfills. The current study demonstrated a sound and efficient approach for characterizing solid waste at the household level. It allowed for the collection of personal data from each of the selected household and therefore represents more reliable results, unlike transfer stations or landfill which represent waste samples from multiple municipal waste sources. It also helped to identify social factors which influence waste generation at household level in informal settlements, this would be a difficult task when sampling takes place at the transfer station or landfill site as there could be mixed waste from different households with different social status.

In conclusion, the waste characterization study on household level provides more detailed, accurate and crucial information on waste composition and the actual process of household

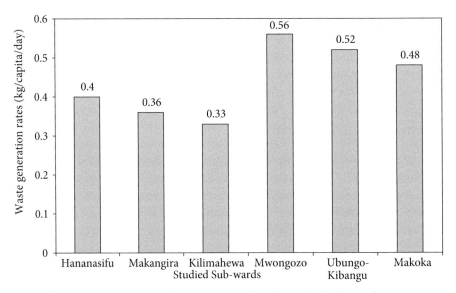

Figure 9.1. The average per capita daily waste generation for studied sub-wards.

waste generation. In combination with the information on the factors governing the generation of waste in the studied sub-wards, it is possible to identify relevant factors to be considered when organizing future interventions in SWM practices at the household level.

9.3.2 The composition of domestic wastes

On average kitchen/food waste stand out as the major (70%) household waste component. We observed that the fractions which constituted food waste originated not only from the normal preparation of households own meals, but also from food related businesses which are conducted at households. In line with this, our survey indicated that 52% of the interviewed households carry out food related businesses in their households. The high presence of food wastes reflects the eating habit and the nature of income generating activities of households which influences the composition of household wastes. This observation can justify the linkage of waste composition and economic activities at household level. The nature of food wastes being organic suggests the possibilities of recycling through composting. Nevertheless, composting as yet turns out to be hardly developed at all within informal settlements. It's potential is something that requires further examination.

Paper accounted for approximately 8% and plastics for approximately 8% of the total household waste as well. Aluminium accounted for approximately 1%, metals approximately 0.8% and glass approximately 1%. Residual wastes accounted for approximately 8%. We found that paper and plastic waste materials are discarded at household level after being used as packing materials for snacks, food and other commodities. This particular observation reflects that these materials are imported from another waste stream and disposed of at household level. So it is likely that household waste contain proportions of institutional or commercial waste. This is similar to the observation that food wastes are generated from business activities taking place within the household environment, however, in contrast for the case of paper and plastics waste comes from elsewhere and find its destination at household level.

The low percentages of metal and aluminium suggest that these waste components are insignificant. Furthermore, the observed practice of removing recyclables such as metal and aluminium amongst others from the waste stream may contribute to this low content. As reflected in Chapter 6 of this thesis, households can remove and keep separate recyclables for their own (re) use or for handing it over to others in return for money or services. The residual waste stems from floor and yard sweepings activities and contains as well ashes from the charcoal and firewood. This reflects the lifestyle and daily household social activities. In our survey, as mentioned in Chapter 6, 76.5% of households discard floor and yard sweepings with other wastes in the same solid waste container. It has been estimated that a large percentage of the population in Dar es Salaam uses charcoal as source of energy[32]. This observation indicates correspondingly high charcoal ash content as a by-product which is discarded as waste.

---

[32] A study by Malimbwi and Zahabu (2008), estimated that charcoal is consumed by 94% of the households either alone or mixed with other fuels. Only 6% of the households do not use charcoal. About 78% of households in Dar es Salaam city use charcoal as their first choice of energy source.

In conclusion, the study on the composition of household waste provided information/data which will help to recognize material categories and their relative proportions in the household waste stream for potential reduction, composting and recycling. In addition, the waste characterization study helped us to identify sources of different waste components and the factors that contribute to their generation. The factors identified were found to correlate to lifestyle related activities such as daily cleaning routines, daily economic activities of households, recovery of recyclable materials within the household, cooking and eating habits.

9.3.3 Relationship between demographic variables and waste characteristics (per capita waste generation and waste composition)

Household size and income of the household are the two demographic variables which are generally considered to be most important factors affecting per capita waste generation and the composition of waste. The bivariate analysis using Pearson's coefficient (r) was employed to see whether there is a correlation between these variables and the per capita daily waste generation.

Household size as unreliable determinant of waste volumes

The mean household size was found to be 5.82 persons calculated from the average household of each studied sub-ward. The bivariate analysis using Pearson's correlation coefficient (r) statistical measure was employed to find the correlation between household size and per capita daily waste. As mentioned in section 5.5.1, it was found that there was no relationship between household size and per capital waste generation. This was justified by the fact that the correlation coefficient of 0.40 obtained from bivariate analysis is very small, indicating small effect of household size on per capita daily waste generation. In other words, the amount of waste generated per person in a household was found not to depend on the number of persons living in the household. This means that household size is not the only factor relating with the daily per capita waste generation in the context of this research. The obvious reason is that some of the household waste fractions are independent of the household size for example waste from the household based business economic activities and residual wastes such as sweepings. In addition, per capita daily waste generation may be independent to household's size as there were variations of household size during the research period. The household size could vary from one day to the other as relatives and friends move in and out. These findings suggest that there is a need to move beyond household size as a quick and easy indicator for domestic waste volumes and instead to examine other both demographic and social factors that underlie the generation of household wastes.

Income of households as predictor of waste volumes

The bivariate analysis using Pearson's correlation coefficient showed that there is a positive correlation (r=0.379) between households' monthly income and per capita waste generation in the studied sub-wards. We thus reject the null hypothesis, that there was no relationship between income of the household and per capita daily waste generation. This means that income and the per capita daily waste generation were found to be two dependent variables in this particular research.

This observation can be supported by the fact that, the monthly income of the household did determine the per capita daily waste generation or the amount of waste produced by a person in a particular household. For that reason, we could argue that income was one of the determinant factors, however, there were other factors which influence waste generation at household level as well. As we earlier reported, various businesses were taking place at the household level, therefore the reliability of the net income given by respondents might be questionable. In addition, a household may not be able to give reliable information because its income was estimated from expenditures per month. Taking into consideration that a large percentage (51.5% of the total) of the studied household were self-employed or working in informal sectors with irregular or variable incomes, it may be difficult to give the proper estimation of their monthly income. The businesses taking place at the household level influence waste generation as solid waste resulting from these businesses are discarded by individuals in the household together with their (other) domestic wastes.

Based on these findings we may argue that a detailed analysis of households' income has yet to be carried out in order to be able to establish the correct relationship between income and household waste generation. Such a precise picture on how socio-economic factors can be linked to (the improvement of) solid waste management at the household levels in the context of informal settlements could be recommended for future study.

9.4 Households as waste handlers

This section answers the second main research question which was addressed in Chapter 6 of this thesis: What are the key domestic practices and routines for handling solid wastes in the primary phase, at the household level? The question was answered by looking at the current waste management practices applied by households, the ways in which domestic solid waste flows 'travel' from the households to the transfer stations, the refuge to alternative disposal methods by households, and the roles of different household members in the practice of waste management. The data needed to answer these research questions were obtained through a household survey (n=360), interviews with 3 solid waste contractors in possession of legal contracts for providing waste collection and disposal services to households, and direct observations carried out by the author.

9.4.1 Solid waste management practices applied by households

Resource recovery

The study showed that households retain some of the items that they do not treat as waste. It was found that 71% of the respondents re-use recyclables and 7% give them to others who might need it, while 22% mix them with other wastes. This practice means that households engage in waste sorting by obtaining recyclables such as plastic and glass containers. This waste management practice exhibits the phenomenon of resource recovery, since this is normally based on the sorting of waste. At the same time, households gave various reasons of not sorting waste when the question was posed to them to explain the reason for not separating waste. The following observations can be made. First, households practice waste sorting in the absence of their knowledge and possibly without recognition by the municipal officials. Second, households sort out the materials which they find they can use for various purposes such as storing oil, medicine, etc. without further knowledge of the principles and purposes of waste sorting as defined from an environmental protection point of view. In other words, this shows that households know the usefulness of the recyclable materials, otherwise they would be burning or disposing them with other types of waste. Their practices of separating and recycling are motivated rather by economic (scarcity) factors than by environmental considerations. Although organic, compostable waste was found to be the by far largest fraction of domestic solid wastes, practices of recycling and re-use of organic waste in terms of (local) composting schemes are not reported from the data.

Although the practice of resource recovery is not recognized by the municipality and organic waste separation not supported and also not common knowledge among householders, waste separation for both technical and economic reasons seems to be an important aspect to be (re) considered in the context of future policies on domestic solid waste management in informal settlements.

Storage of domestic wastes

Waste storage involves placing the wastes that are generated in waste storage containers for collection. The study showed that different types of waste containers with different sizes are used by households as storage facilities as shown in Figure 9.2.

17.5% of respondents use old plastic buckets, while 73.5% uses plastic bags, 7% uses boxes of various sizes, and the remaining 2% does not have any waste container. Waste containers are considered as facilities without any value to households, and therefore perceived as waste too. This

Figure 9.2. Different types of waste storage containers used by households under study (photo by the author, 2008).

is reflected by the practice of households using waste storage containers that they eventually dispose of together with the wastes. This emphasizes that households understand the adverse effects of being surrounded by wastes. They want a clean environment, as it was expressed during the research in several ways. For example plastic bags that they use to carry commodities from shops and markets; very old buckets which cannot be used for any other purpose and boxes. Although Kinondoni Municipal waste management by-laws[33] obliges households to have standard storage containers, the Kinondoni and City health officers admitted that the majority of people in Kinondoni and Dar es Salaam city as a whole are not aware of the existing by-laws. Surprisingly, the officers who are in charge of enforcing the by-laws were admitting the shortcomings. The reasons for this lack of implementation are the difficulties in enforcing them in informal settlements due to existing living conditions such as congestion, low security and low earnings. If households are not aware of the by-laws that can be one of the reasons of having freedom of choosing their own means of storage facilities which are easy for them to handle in terms of maintenance, and those which suit their living environment. For instance, given the fact that the nature of their living environment is an informal settlement where the security is very low, placing a plastic waste container outside the house means running the risk of the official container being stolen. Respondents, therefore, prefer to use containers which have no value even to thieves. That is why they use plastic bags or very old buckets as shown in Figure 9.2. From this understanding, we can conclude that, however, a good by-law might be, if it not applicable to the situation on the ground becomes irrelevant which will not guarantee the proper waste management practices by households. What we do want to conclude here is that social prerequisite is relevant when it comes to improve household solid waste management.

The practice of placing waste containers in the yards of the house has a cultural background. For most Africans it is unacceptable to store waste inside the house where people sleep. This finding is similar to the observation made by Addo-Yobo and Ali (2003) in Ghana. They underlined that the practice of storing waste bins outside the main house or in places where visitors cannot see waste bins is socially rooted.

As our empirical findings revealed, at the household level, SWM is considered to be a housekeeping duty, and so it is perceived as the domain of housemaids and children under the supervision of women. Storage of waste at the household level is usually considered and carried out as part of the households' chores. Thus, the main responsibility for SWM within the household falls on the women's shoulders; they are expected to execute and supervise the household work and to take care of the solid waste. This implies that solid waste management is given a low status at household level with male adults not taking any responsibility. As Anand (1999) pointed out the basic attitude towards waste management as a menial task and a low status job implies that even within a household this task is likely to be done by the weaker or low-status members, for instance, children or housemaids.

---

[33] Municipal by-laws define responsibilities for households such as the obligation to have two solid waste collection receptacles (one for organic and the other for non-organic waste) of not less than 40 l fitted with a lid (Collection and Disposal of Refuse By-law, 2001 Section 4 (1 and 2) and Section 5). These by-laws also require the occupiers of premises to maintain receptacles to keep waste and prohibit people from causing a nuisance and depositing waste on streets or in open spaces not designated as collection points.

These findings acknowledge the existing differences within households between household members (men, women, children, housemaids) with respect to the waste management roles. Based on these findings, the most important point to be communicated is that these differences should be taken into account when planning for the improvement of SWM at household level. If not taken into account, social and cultural factors as discussed here might turn out to be important constraining factors for realizing more sustainability in domestic solid waste management by households.

Waste flows 'travelling' from the households to the transfer station

In analysing the waste-flows in between households and transfer stations, we made use of the distinction between the primary and the secondary phase of the solid waste management chain. The skip, container or transfer station becomes an important facility in the SWM chain as it not just receives wastes from households but as well because it functions as the point of connection between the primary and secondary phase of the waste chains and its main actors. In the pre-transfer station phase, householders and their everyday life rationalities and practices of storing, (non) separating and delivering wastes are the dominant kind of practices and they should be recognized as such also by actors operating in the secondary phase of the waste chain. The importance of the transfer station is demonstrated by the three ways in which domestic waste flows might 'travel' or be transported from the households to the transfer station. The three modalities of transport are discussed in detail in Chapter 6 of this thesis. The first way involved the transfer of waste by household members, and the second way involved transfer of waste by using waste collectors employed by formal waste contractors. A third way was through illegal waste pickers who pick waste from households to the transfer station. As has been noted in Chapter 6 informal waste pickers are involved at the very first point of collection from households, running parallel with formal stakeholders in their efforts of collecting waste from households. The findings of several studies have indicated that in developing countries there often exists a linkage between formal and informal waste practices in SWM. This include studies of Medina (2000) and Wilson *et al.* (2006) amongst others which have shown that the informal sector plays an important role in SWM systems especially with regard to waste recovery. Our findings contribute further on the previous studies, by revealing that informal waste pickers play an important role with regard to waste collection and disposal at the household level.

In receiving waste from households, transfer stations become a meeting point where different waste chain actors meet or cluster together. As noted in Chapter 6 and Chapter 7 of this thesis, households, formal and informal stakeholders can become connected in many different ways in the process of collecting and transporting domestic wastes from the households to the transfer station. In the process of collecting and transferring domestic wastes, we identified the following forms of integration or clustering: (1) households and formal stakeholders, (2) households and informal stakeholders, and (3) households, formal and informal stakeholders and the transfer station itself. As seen from the three clusters of integration, households appear in each cluster or set of the integration. This illustrates that households can be seen to play the most crucial role in SWM practices before the transfer station. Householders are the binding factors that integrate formal and informal waste management systems at the transfer station. On the other hand, waste management practices by households and service providers (formal and informal stakeholders) are accommodated or integrated around the transfer station. With this observation, we aim to illustrate the relevance of the theory of MMA in particular with respect to its claim that the involvement of end-users (this case households) in solid waste management practices is important and has to be researched in close connection with the technologies and infrastructures, as represented in this case by the local transfer stations for domestic solid waste management. These transfer stations become an important location or focal point from households' waste management perspectives.

9.4.2 Management of waste transfer stations

Waste contractors – under the supervision of the local municipal authorities – are responsible for the management, supervision and operation of the transfer station. Identifying the exact placement or location of transfer stations turns out to be an important item and was decided upon by households, waste contractors and the local leaders of the respective sub-wards. However, identifying a suitable site for placement of a waste transfer station can be challenging question because it has to be accessible both for households as well as for actors operating in the secondary phase of the waste chain. In this study as we noted in Chapter 6, transfer stations are stationed at a point along the street within the sub-ward where households can access. This shows that in locating the transfer station, the consideration of accessibility is taken into account by households. Another crucial factor as far as the location is concerned is the distance between the households and transfer station, because the aim is that all households can walk easily to reach the transfer station. According to Schübeler *et al.* (1996) for householders to carry their own waste to a transfer station this needs to be located within easy walking distance to discourage indiscriminate dumping. He recommended a walking distance of 100 metres. The study of Parrot *et al.* (2009) noted that the distance between houses and garbage bins affects domestic waste disposal behaviour. He found that less than 50% of his studied households use communal bins which were located about 1,600 m. He concluded that that long distances explain why households dispose domestic wastes in open areas. Within the scope of this study, it was difficult to measure exactly the distance which households travel to dump waste in transfer stations. The information obtained from the interviews with the service providers as far as the furthest distance travelled by households to reach the transfer station ranged between 200 to 400 metres. However, these findings cannot be taken to be very reliable because they are based on estimations without using any metric measurements. Also the study did not go into details finding out the number of households delivering waste to transfer station.

However, the most important thing to note here is that households take part in deciding about the location of the transfer station. In other words we can say that households have a say in the location of transfer station where they transfer their waste. This way of including households by consulting or informing them about the location of transfer station, provides an opportunity to come up with optimal solutions which will encourage households to bring waste to transfer station and minimize open dumping. Since households are the ones using the system, we can speculate that they have knowledge on the local situation which will take into consideration the distance and accessibility to the location of transfer station decided upon by them. When giving householders a say in the location of the transfer station, they are perhaps prevented to resort to other, more unsustainable and unsafe methods of getting rid of their wastes.

9.4.3 Alternative disposal methods

As a result of not having access to any formal waste collection service, households in Makoka, Mwongozo and Ubungo-Kibangu sub-wards are compelled to find other means of getting rid of their waste. Burning, burying and illegal dumping practices are regularly used options for these households to dispose of their wastes. These practices are commonly conducted by household members or informal waste pickers in nearby streets or in the open space. Based on the survey, respondents of these sub-wards showed their concern over their own practices and claimed that the municipality should provide a waste contractor to serve them. The concern of these households is an indication that they are aware of the effects of bad management of disposing domestic wastes. These practices show that, besides the reason of not having any formal waste collection and disposal services, waste is not wanted; and households want to get rid of it immediately and effectively.

With respect to our findings, households in Hananasifu, Makangira and Kilimahewa sub-wards practice burning, burying and illegal dumping of wastes despite of having access to formal waste collection services. In our survey, respondents indicated that, informal waste collection practices are more reliable and affordable as compared to the formal collection services. Possibly, the reliability and affordability of informal waste collection services made household to value informal services more than the formal services. Reliability of the service is the most crucial element for households irrespective of who is providing waste collection services. From the focus groups it became clear that informal pickers in general have better performance than formal contractors although the overall level of service provision is still seen as inadequate. Although Kinondoni municipal waste management by-law forbids the practices of burning, burying and illegal dumping, and despite the existence of formal and informal waste management services in the sub-wards of Hananasifu, Makangira and Kilimahewa, it is interesting to note that these practices are still prevailing. It is possible that these practices are taken by households as normal in managing waste because households have not been barred by the municipal officials, and probably the regulations by the municipality over these practices are not strong, and households are not held responsible for their practices.

We conclude from our discussion that householders, when confronted with inadequate and unreliable waste collection services, as provided by both formal and informal service providers, develop alternative waste handling options such as to burn, bury or dispose of domestic wastes in unauthorized places. The persistence of these practices can also be attributed to the poor enforcement of regulations from the side of the municipality. This conclusion is in line with the argument put forward by Anand (1999) that, when primary waste collection services are not reliable, the incentive is to explore other options and when regulations is either absent or the majority are non-compliant, the incentive is to dump waste in open access spaces such as streets and public spaces.

9.5 Households as service recipients in solid waste management chain

This section answers research question 3 which was addressed in Chapter 7: 'How do households receive solid waste management services and what are their basic perceptions and evaluations of the ways of being served?' In Chapter 7 we explored in detail the relationships householders uphold with other – both formal and informal – actors in the solid waste management chain, the kind of institutional support which exists for these relationships, and the ways in which households evaluate and assess the formal and informal relationship they uphold with other stakeholders?

9.5.1 *The relationships of households with other stakeholders in domestic solid waste management*

Households as recipients of solid waste service become embedded in SWM chains through both formal or informal relationship. The formal links are those of the households with waste contractors and the municipality. As such the formal relationships are supported by the waste management structure of the municipality and include stakeholders who enjoy official recognition and protection; thus in this way the roles and responsibilities are well defined in the municipal by-laws.

Households are in the weaker position of the structure. As passive recipients of services, they are placed at the down-stream or bottom-end of the chain, with the municipality being at the top, assigned with the formal powers to organize, monitor and supervise adequate service delivery. In this formal structure, the municipality is recognized as the key service regulator, whereas waste contractors are recognized as actual and formal service providers when registered and given a license by the municipality. The responsibility of the municipality to the households is to make sure that they receive the services of waste management in order to avoid the outbreak of the diseases and to protect the environment in general.

This is the situation as it exists on paper. And during the focus groups and in the survey the householders expressed their awareness about this formal structure. They all held the municipality responsible for waste service delivery. The actual situation of being underserviced or not serviced at all results in a basically critical attitude towards the official authorities.

With regard to the informal relationships householders uphold, we identified and discussed two forms in Chapter 7. First, there is an informal relationship between households and illegal waste pickers which are unregistered and unregulated, and therefore not officially recognized by municipal officials. Second, informal relationships can exist between households and waste contractors as they result from the mutual agreements between households and waste contractors on issues which are not covered by the municipal by-laws. Both the formal and informal relationship exist parallel, next, complementary and/or in competition with each other in the context of the formal SWM system. Because of their omnipresence in informal settlements, formal and informal relationships are two relevant dimensions of domestic solid waste management systems and they have to be taken into account when improving SWM-services at the household level in informal settlements. It is worth remembering that in Kinondoni municipality more than 60% of the population live in informal setting, outside the legal system and its official parameters as stipulated by urban planners.

9.5.2 The role of municipal authorities in assisting stakeholders in the waste chain: the role of public-private parnerships

The Kinondoni municipal council takes its responsibility for supporting household solid waste management actors in the provision of waste collection and disposal services. In our research, we documented the kinds of support that exist, both through the provision of labour forces, legal provisions and equipment. The employment of lower municipal waste management officers to supervise waste management at the household level results in households becoming in direct contact with them. This demonstrates the following: First, it supports the notion of decentralization of power to lower levels of the municipality in order to ease implementation and recognizes the households' role in waste management. In addition, the problems relating to household waste management are communicated to higher authorities in a more official way to influence the implementation and decision making. Second, it promotes the relationship between households and municipal authorities by linking them to higher authority through their local leaders. Local leaders are important to households because they are closest to the people and people trust them because they are elected by people. The trust of households to local leaders is demonstrated by our empirical findings of Chapter 7 where we found that waste contractors use local leaders to convince households to come into agreement on certain SWM issues such as payments, modes of use and location of transfer stations, etc. The local (*Mtaa*) leaders are the ones selecting franchisee/contractors to provide waste collection service in their areas of jurisdiction.

In the context of these efforts to bring local authorities and service provision closer to households the promotion of Public-Private Partnerships (PPP) is assigned an important role. The municipality engages in the contracting out of waste services provision under a franchise[34] system. According to Nkya (2000), Public-Private Partnership is viewed, as a means to improve the provisioning of public goods and services in terms of enhancing effectiveness, efficiency, adequacy and responsiveness. In the case of solid waste management, the ultimate aim of PPP is to provide effective SWM services to residents in terms of the amounts of waste collected, and the sustainability of the service provisioning. It is important to note that Nkya's study focused on the Public-Private Partnerships established in the area of solid waste management between Dar es Salaam City council and the private sector. Interestingly, looking at the role of the municipality in assisting stakeholders in the current study, two cases of PPP can be noted. The first case is the PPP between the municipality and waste contractors; while the second case of PPP is between the waste contractors and households (service recipients). The first case of PPP is governed mainly by a legal contract that defines the partnership and specifies the roles and responsibilities of the partners (municipality and waste contractors) involved. Of the importance in this case is that, the legal contract is the pillar of the municipality to monitor the performance of the waste contractors and to enforce sanction if necessary. The second case of PPP found in our study is not legally recognized. This type of PPP emerged from the joint efforts of households, waste contractors and local leaders in managing domestic wastes. The joint effort has been possible

---

[34] Government grants a private firm an exclusive monopoly to provide a specific type of solid waste service within a specific zone. The firm collects its own revenues from generators within the zone or from the sale of solid waste by-products removed from the zone.

due to the institutional set up of the municipality which enables households to be closer to local leaders and waste contractors. Therefore, it can be concluded that the municipality has provided the opportunity for both official and less official forms of partnership in domestic solid waste management at the household level.

Another important issue of institutional support is the assistance of the municipality in the form of supplying moveable standby trailers to waste contractors via sub-ward leaders. This has supported the integration of households and informal waste pickers in the SWM chain; as we documented in Chapter 7. Moveable standby trailers are used as transfer station which receives waste collected from households through service providers (formal and informal) and households themselves. Though the municipality stated that they do not support informal waste pickers; the forms of support identified in our research clearly show that indirectly the municipality seeks to influence the integration of informal waste pickers in the SWM chain. This is explained by the fact that, informal waste pickers are serving the system while the municipality has the overall authority. In other words, informal waste pickers are operating in a regulated and organized SWM system.

To conclude this section, the assistance provided by the municipality to stakeholders in managing wastes at the household level can be discussed in relation to the theory of modernized mixtures as used in this research. Within the framework of modernized mixtures the municipality provides labour, technical assistance, and legal support for waste contractors when improving waste collection and disposal services in informal settlements. Under the privatisation regime, the municipality gives out formal contracts for waste collection and disposal to waste contractors. In reality, we see a mix of institutional forms emerging, depending upon both formal and informal relationships. These mixtures found in real life are not formalized or institutionalised in an official, legal way. They represent however very important relations between households and other actors – both formal and informal – in the waste chain. As such, they are the kind of arrangements which make waste management services possible and which aim to facilitate the improvement of waste collection and disposal services in informal settlements.

9.5.3 Households' perceptions and assessments of waste service provisioning

Households perspectives on transfer stations and their role in service provisioning

We documented the perceptions of households in evaluating solid waste management system, in detail in Chapter 7. We used the method of survey research to get to know the opinions of households on transfer stations and related forms of waste service provisioning. From the survey 45.3% of all the studied households indicated they want transfer stations in their neighbourhood, while 35.6% indicated that transfer stations produce unpleasant smells and for that reason are not to be welcomed in the neighbourhood. Both the survey and the focus groups revealed that there is a strong resistance against transfer stations when placed within a short distance of dwelling areas (houses). Similar observations of local residents opposing transfer stations were reported by Alam *et al.* (2008), when local residents in Kathmandu raised objections on the construction of a transfer station and work on this was suspended. In our case, households were against the presence of transfer station in their neighbourhood for reasons of negative externalities such as health hazards, and bad odour. Health hazards resulting from the unattended transfer station were the major concern of these households. For instance, the respondents from Hananasifu sub-ward claimed that they prefer a moveable transfer station because they have had the experience of a permanent transfer station which turned to be a complete nuisance. When this research was being undertaken, households in Hananasifu sub-ward were bringing their waste to a moveable transfer station which was not left to stay in the neighbourhood for a longer time. With this experience, the major concern of households was for the municipality to provide enough moveable transfer stations to the waste contractors. This perception shows that households want their surroundings to be as clean as possible and without any waste problems.

On the other hand, a majority of interviewees, 61% from Makoka, Mwongozo and Ubungo-Kibangu sub-wards indicated that it is important to have a transfer station in their neighbourhood. This response might be due to the problems they are facing of not having any formal solid waste collection and disposal services, so they are in favour of having transfer stations where they can dump their waste. In spite of the fact that they want the municipality to provide waste transfer stations, they want the transfer station to be located somewhere not very close to their neighbourhood. Implication of households perceptions towards the location of the transfer stations is that, their perceptions and opinions should influence the decision making process about the location of transfer stations. Essentially, the decision on the location of transfer station should take into consideration the extent of resistance by households against the location of the transfer station in their neighbourhood in addition to the factors such as health impact and accessibility. The transfer station serves as an important object in the efforts of householders to claim more (co)decision-making powers and to increase levels of participation

Regarding the assessment of households with respect to the quality and nature of waste service provisioning, our study documented a persistent orientation of householders on the official authorities. Our findings indicated that most households (52.8%) hold the municipality responsible for waste service delivery. In other words, households thought that SWM services are the sole or at least the prime obligatory responsibility of the municipal authorities. However, as explained in Chapter 2 of this thesis, the municipal authorities in general failed to provide the service to the city population themselves in an efficient way. For that reason, they opted for a stronger role of the private sector. This involvement of private waste contractors however did not result in any significant improvement in the levels of service provision. Informal waste pickers stepped in to make up for failed and irregular nature of the service provision by formal, private actors.

It is against this background that we have to interpret the findings that the majority of the households hold the municipal authorities responsible for inadequate service provision. Households tend to blame the municipal authorities and do not consider themselves to be first or primary responsible for improper, unsafe and unsustainable forms of domestic solid waste management.

This study focused on households and their perceptions and assessments of waste management services. It would be important in future studies to also capture in more detail the perspectives of the service providers, so as to get a combination of perspectives which might contribute to finding better solutions for SWM in informal settlements.

Households perspective on their own roles in solid waste management

In Chapter 6 we discussed the perceptions of households about their own roles in solid waste management in more detail. We found that 80.6% of the respondents perceived that they would pay for the service if sufficient services were provided to them. This perception reflects that households are not particularly feeling that they have a role to play; rather they have set a precondition of having a reliable waste collection and disposal services set in place for them by others. If everything is arranged properly, they intend to pay for the service. This condition indicates that, households regard themselves as customers or clients in waste management services. They want to pay for reliable services and don't want to actively participate in other SWM activities; even though they are aware (from survey and focus group discussions) of the health risks resulting from uncollected or scattered wastes around their premises. They more or less take a passive role as receivers of waste services and don't think about a pro-active, initiating role for householders themselves. Households are principally concerned with the (health and environmental) consequences of mismanagement of domestic wastes; but all they want is to have their wastes removed from their premises. Or, in other words, households are mainly interested in receiving effective and dependable waste collection and disposal services within their immediate vicinity. They do not bother about where the waste goes or how it is dealt with. Their ultimate concern is a clean living environment within their immediate premises. They expect the municipality to perform its duty on SWM and keep their environment clean.

9.6 Reflection on the conceptual framework: flexibility, accessibility and ecological sustainability

As mentioned in the introductory section of this chapter the performance of a specific 'modernized mixture' in domestic solid waste management is to be judged against a set of three criteria: ecological sustainability, accessibility and flexibility. These criteria are reflected in our conceptual framework in a combined way. From our empirical findings we conclude that the two criteria of flexibility and accessibility have been more important and central to the research project when compared to the ecological sustainability.

The practical approaches used in waste collection and disposal to some extent show the features of flexibility, specifically in terms of emerging partnerships and when it comes to institutional arrangements. Institutional arrangements and partnerships have allowed responsibilities of solid waste services provision to be shared among the households, waste contractors and the municipality in ways that are governed by local circumstances instead of by official law and regulation. The authority that is granted to the municipality to formulate by-laws allowed for the privatization of solid waste collection and disposal to start playing a role. In this way, the municipality is still responsible but in a more indirect way, by providing an enabling framework for the service providers and to support their activities with equipment, legislation, and monitoring and law enforcement. Since privatization is supported by the central government this enabling framework – including the by-laws and the forms of institutional support in place – ensures the stability of the waste management system, also in case of political and economic instability. Mixtures of public and private arrangements, of formal and informal relationships contribute to the flexibility and resilience of the systems, so it might be argued. The criterion of accessibility is of course among the most crucial aspects of solid waste management services and infrastructures in informal settlements. This criterion can be about social accessibility, when dealing with payments and fees for services and when discussing the walking distances to transfer stations. This criterion can refer as well to the technical and physical aspects of the waste system, when discussing the accessibility of areas for both formal and informal service providers. Large trucks perform less than handcarts and wheelbarrows in densely populated areas with small, unpaved roads.

Flexibility and accessibility criteria are also at play when analysing the role of informal and illegal waste pickers who offer their services to households who are not being served. By functioning as back-ups for or complementary to the formal system, the informal waste pickers contribute to the accessibility and flexibility of the waste management system. In conclusion, the criteria of flexibility and accessibility are reflected to a large extent by our empirical results, more so than ecological sustainability.

9.7 Main conclusion

Households generate wastes, also when they are located in one of the socially dense neighbourhoods of the informal settlements in the capital cities in East Africa. These wastes are the result of the various domestic social and economic activities, and their volume and composition are shown to depend not so much on the (unpredictable since fluctuating) sizes and incomes of the households but more on the kind of (small, informal) business activities the members of the household are engaged in.

Once generated, the wastes have to be stored somehow and somewhere until the moment the wastes can be handed over to a third party that will take care of the wastes. Also, during and shortly after the process of generation, different waste fractions or resource components can be separated because they still represent economic value for the households or because the separation makes possible a more effective disposal, treatment or reuse of the wastes. In our study it was shown that in informal settlements like Kinondoni, waste storage is done with the help of 'useless' plastic bags and containers which are disposed of together with the waste. Except for a few valuable components, waste separation is not a very common practice. Given the fact that around 70% of all the domestic wastes are organic in nature, it may come as a surprise for environmental researchers that waste separation for composting does not seem a common practice at all in informal settlements.

When wastes are not properly disposed of and timely removed from the living environment of the households, a lot of negative consequences are shown to result. Health risks, nuisance, bad smells, degradation of eco- and water-systems are all recognized by the householders in our research as the unwanted, negative side effects of ill functioning waste management systems. And mal-functioning, mismanagement, non-servicing and underservicing are more appropriate terms to refer to the everyday reality of waste management in informal settlements than effective and efficient separation, storage, removal and disposal of domestic solid wastes. Underperformance by almost all actors involved seems closer to being the rule instead of the exception.

This PhD-research can be read as a search for answers on the question how to understand the origins and the reasons behind the problems in SWM at the household levels in informal settlements. Who is to blame and for what reasons? And more importantly, how can and should the situation be improved. Who has to take the lead and what should be done first? At the end of this research, we have to admit that simple answers are not available. There is not one party to blame, and there is not one single most important factor or strategy available to solve these problems in the short run. Is there nothing to conclude then and are there no lessons at all that can be learned from this extensive research project? Of course there are lessons learned, and we offer them in the form of four concluding considerations we think to be of relevance for solid waste management in informal settlements.

1. Both technical and social factors play a role.

    Lack of proper equipment, lack of human resources, inaccessibility of particular areas for trucks, lack of space for the storage of waste, absence of proper (storage, separation, removal) technologies etc. are all very visible and relevant factors to explain the malfunctioning of solid waste management systems in informal settlements. However, as this study has shown, these technical and infrastructural factors cannot explain by themselves the waste situations as they came in existence and tend to persist in the informal settlements in Dar es Salaam. Past performance of waste actors, lack of income in poor households, established habits of 'alternative waste handling', cultural norms on cleanliness, divisions of roles within households, and the low social value of waste handling practices are all examples of social factors that co-determine failure or success of waste handling systems

2. Both formal and informal relationships are important.

    When formal waste management systems do not deliver services at an adequate level, informal practices, actors and relations tend to come into play. And once they are at play, they tend to develop into established practices with stakeholders and specific interests attached to them. The practices of waste burning, burying and dumping as developed by households in response to the absence of regular waste collection and disposal services tend to gradually develop into socially accepted routines for handling wastes. The activities of illegal waste pickers solving the immediate waste problem for households on a cash-for-delivery basis tend to develop into normal arrangements appreciated by the parties directly involved. As our study has shown, informal relations and activities are indispensable elements of the waste management systems in informal settlements and need to be recognized and dealt with when trying to improve the functioning of these systems in the short run.

3. Households as key actors should be recognized for doing things their own way.

    This research is unique in its strong focus on the role of households in solid waste management practices in informal settlements. Their activities of generating, storing, sorting, separating, transporting, (illegal) disposing of domestic wastes should be investigated, analysed and properly understood by all other parties involved in domestic solid waste management. This study documented a disparity to exist between the municipal requirements on the one hand and the actual practices of household in the SWM chain on the other. For instance in waste storage, household use containers of their own choice whereas the municipality has different formal requirements. In resource recovery, households were found to practice resource recovery by setting wastes aside (without putting the recyclable materials in a special container) while the municipality specified that all households are required to have two solid waste collection receptacles (one for organic and the other for non-organic waste) of not less than 40 l fitted with a lid. In practices of waste transfer, households use the system of bringing waste to the transfer station following their own agreement with the waste contractors and paying 'on the spot', whereas according to municipal by-law the households are required to pay waste collection fees, etc. What is documented here is an existing mismatch between the (legislative and system) rationality of municipal authorities on the one hand and the social and life world rationalities of householders on the other. An important lesson learned from this research is that, in order to effectively improve solid waste management at the local level of households, the specific nature and dynamic of household practices have to be considered and taken into account when designing new interventions, both socially and technically.

4. Solutions are not one of a kind but 'mixtures' that start from local circumstances.
    The conceptual model used in this research was based on the theory of the Modernized Mixtures as developed in the context of the PROVIDE project. This theory suggests that systems and infrastructures for the handling of waste-water and solid wastes should be adapted to the local situation both in physical and social respects. Instead of bringing a fixed set of socio-technical elements into a community or a city in a pre-defined way, the theory suggests to develop infrastructures and their corresponding management structures bottom-up, in dialogue with the end-users of the infrastructures and services. The present research has shown that the situation of structural underperforming waste management systems in informal settlements in Dar es Salaam represents a complex, multidimensional problem that cannot be solved overnight or by one particular actor or strategy and that a mixture of actors and strategies is required.

9.8 Some suggestions for future research

Based on the empirical findings and conclusions we have identified a number of limitations of the present research that can be translated in suggestions for future investigations on this topic.

The study showed that factors relating to social and economic activities were highlighted to have great influence in waste generation at household level. The use of household premises for daily social and business activities may imply a waste generated at household level in relation to those activities; hence there is a need for further study to establish more detailed information on household waste generation and composition as well as the relationship between social-economic activities and household waste generation rates.

Further research is needed to investigate in more detail the mismatch between the formal image of the waste system as described and prescribe by municipal laws on the one hand and the everyday waste handling practices adopted by households in informal settlements at the other. How can the gap between image and reality be bridged? How can waste management policies become more adapted to the social, economic, and physical circumstances that are characteristic of the local situation of informal settlements? We think the role of participation, transparency and organized feed-back to householders deserve further investigation in this respect.

Further study which might benefit policy makers relates to clarifying the link between waste handling practices of the households and their perceptions and assessment of the existing waste management system. How do the willingness to participate in and pay for waste services relate to householders' opinions about the (lack of) performance of the other stakeholders, both formal and informal, in the waste management chain? How is the perception of their own role related to their expectations and opinions with respect to the roles to be performed by others.

Informal relations turned out to play an important but not officially recognized role in solid waste management in the primary phase of the waste chain. More detailed studies are needed to investigate the actual and potential contribution of informal relationships and in particular the role of informal waste pickers.

Finally, the geographical scope of this kind of studies needs to be broadened. This research studied households living in informal settlements in Dar es Salaam city through quantitative and qualitative methods, while in Nairobi and Kampala the study was limited to qualitative information. As a result, the comparative analysis offered in this research is limited in its scope and ambition. Future research on the role of households in solid waste management practices at the local level of informal settlements should take a full, mature comparative perspective between a range of local settings which are all research with a similar set of methodologies.

References

Abduli, M.A., 2007. *National Reports: Islamic Repbulic of Iran Solid Waste Management: Issues and Challenges in Asia.* Tokyo, Japan: Asian Productivity Organization, pp. 92-117.

Abu Qdais, H.A., Hamoda, M.F. and Newham, J., 1997. Analysis of Residential Solid waste at Generation Sites. *Waste Management and Research*, 15, 395-406.

Achankeng, E., 2003. Globalisation, Urbanisation and Municipal Solid Waste Management in Africa African Studies Association of Australasia and the Pacific 2003 Conference Proceedings – African on a Global Stage, University of Adelaide, p. 22.

Addo-Yobo, F.N. and Njiru, C., 2006. Role of Consumer Behaviour Studies in Improving Water Supply Delivery to the Urban Poor. *Water Policy*, 8 (2), 111-126.

Addo-Yobo, F. and Ali, M., 2003. Households: Passive Users or Active Managers? The Case of Solid Waste Management in Accra, Ghana. *IDPR*, 4 (25), 373-389.

Adedipe, N.O., Sridhar, M.K.C., Baker, J. and M.V., 2005. Waste Management, Processing, and Detoxification. In: K. Chopra (ed.), *Ecosystem and Human Well-being: Policy Response Volume 3.* Washington, DC, USA: Island Press, pp. 313-355.

African Development Bank, 2002. *Study on Solid Waste Management Options for Africa. Project Report: Final Draft Version.* Abidjan, Côte d'Ivoire.

Ahmed, A.S. and Ali, M.S., 2006. People as Partners: Facilitating People's Participation in Public-Private Partnerships for Solid Waste Management. *Habitat International*, 30 (4), 781-796.

Al-Khatib, I.A., Monou, M., Abu Zahra, A.S.F., Shaheen, H.Q. and Kassinos, D., 2010. Solid Waste Characterization, Quantification and Management Practices in Developing Countries. A case study: Nablus District, Palestine. *Journal of Environmental Management*, 91, 1131-1138.

Alam, R., Chowdhury, M.A.I., Hasan, G.M.J., Karanjit, B. and Shrestha, L.R., 2008. Generation, Storage, Collection and Transportation of Municipal Solid Waste – A Case Study in the City of Kathmandu, Capital of Nepal. *Waste Management*, 28, 1088-1097.

Ali, M. and Snel, M., 1999. *WELL Study. Lessons from Community-Based Initiatives in Solid Waste: Task No. 99* Loughborough, UK: London School of Hygiene & Tropical Medicine, WEDC, Loughborough University.

Anand, P.B., 1999. Waste Management in Madras Revisited. *Environment and Urbanization*, 11 (2), 161-177.

Anschütz, J., 1996. *Community-Based Solid Waste Management and Water Supply Projects. Problems and Solutions Compared. A Survey of the Literature.* Gouda, the Netherlands: WASTE.

Anschütz, J., IJgosse, J. and Scheinberg, A., 2004. Putting Integrated Sustainable Waste Management into Practice Using the ISWM Assessment Methodology: ISWM Methodology as Applied in the UWEP Plus Programme (2001-2003). Gouda, the Netherlands: WASTE, pp. 42-46.

Bakker, S., Kirango, J. and Van der Ree, K., 2000. *Both Sides of the Bridge: Public-Private Partnership for Sustainable Employment Creation in Waste Management Dar es Salaam, Planning for Sustainable and Intergrated Solid Waste Management.* Paper presented at the Workshop in Manila, the Philippines, pp. 15.

Bandara, N.J.G.J., Wirasinghe, J.P.A.H.S.C. and Pilapiiya, S., 2007. Relation of Waste generation and Composition to Socio-economic Factors: A Case study. *Environ Monit Assess*, 135, 31-39.

Barr, S., 2007. Factors Influencing Environmental Attitudes and Behaviors. A U.K. Case Study of Household Waste Management. *Environment and Behavior*, 39 (4), 435-473.

Barr, S., Gilg, A.W. and Ford, N.J., 2001. A Conceptual Framework for Understanding and Analyzing Attitudes Towards Household-Waste Management. *Environment and Planning*, 33 (11), 2025-2048.

Baud, I., 2004. Markets, Partnership and Sustainable Development in Solid Waste Management; Raising the Questions. In: I. Baud, J. Post and C. Furedy (eds.), *Solid Waste Management and Recycling: Actors, Partnerships and Policies in Hyderabad, India and Nairobi, Kenya*. Dordecht, The Netherlands: Kluwer Publishers, pp. 1-18.

Baud, I. and Dhanalakshmi, R., 2007. Governance in Urban Environmental Management: Comparing Accountability and Performance in Multi-stakeholder Arrangements in South India. *Cities*, 24 (2), 133-147.

Baud, I., Grafakos, S., Hordijk, M. and Post, J., 2001. Quality of Life and Alliances in Solid Waste Management: Contribution to Sustainable Development. *Cities*, 18 (1), 3-12.

Baxter, P. and Jack, S., 2008. Qualitative Case Study Methodology. Study Design and Implementation for Novice Researchers. *The Qualitative Report*, 13 (4), 544-559.

Bolaane, B. and Ali, M., 2004. Sampling Household Waste at Source: Lessons Learnt in Gaborone. *Waste Management and Research*, 22, 142-148.

Burian, B., 2000. *Sustainable Development in an Urban Tanzania Context*. Cities of the South: Sustainable for Whom, 3-6 May, N-AERUS International Workshop. Geneva, Switzerland.

Chinamo, E.B., 2003. *An Overview of Solid Waste Management and how Solid Waste Collection Benefits the Poor in the City of Dar es Salaam*. A Paper Presented at the International Workshop on Solid Waste Collection that Benefits the Poor.

Chung, S.S. and Poon, C.S., 1995. The attitudinal differences in source separation and waste reduction between the public and the housewives in Hong Kong. *Environmental Management*, 48, 215-227.

Cointreau-Levine, S.J., 1994. *Private Sector Participation in Municipal Solid Waste Services in Developing Countries*. Washington, DC, USA: World Bank.

Cointreau, S., 1982. *Environmental Management of Urban Solid Waste in Developing Countries*. Washington, DC, USA: The World Bank.

Commonwealth, 2003. Public Private Partnerships: A Review with Special Reference to Local Government. In: Anonymous (ed.) *Commonwealth Local Government Handbook*. Rochester, UK: KPL.

CWG-WASH, 2006. *Solid Waste, Health and the Milenum Development Goals*. A Report of the CWG International Workshop. Kolkata, India.

Dahlen, L., 2005. *To Evaluate Source Sorting Programs in Household Waste Collection Systems*. Luleå, Sweden: Luleå University of Technology, 146 pp.

Dastidar, S., Cerdena, A. and Terza, A., 2007. *Project Report: Local Agenda 21 Kampala Better Waste Management for Better Lives*, London, UK: University College London.

DCC, 1993. Proceedings for the City Consultation on Environmental Issues: Dar es Salaam.

DCC, 2004. *Dar es Salaam City Profile*. Dar es Salaam: Dar es Salaam City Council.

Dedehouanou, H., 2004. Organizational Puzzle of Household Solid Waste Management in Porto-Novo. In: Anonymous (ed.) Human Settlement Development. The Central Role of Cities in our Environment's: Future Constraints and Possibilities. EOLSS. Available at: http://www.eolss.net/ViewChapter.aspx?CategoryId=14.

Dharmawan, A.H., 1999. *Farm Household Livelihood Strategy, Multiple Employment, and Socio-economic Changes in Rural Indonesia: Case Studies from West Java and West Kalimantan*. Discussion Paper: Institute of Rural Development, the University of Goettingen.

Drangert, J.O.J.O., Okotto-Okotto, J., Okotto, L.G.O. and Auko, O., 2002. Going small when the city grows big new options for water supply and sanitation in rapidly expanding urban areas. *Water International*, 27 (3), 354-363.

Ekere, W., Mugisha, J. and Drake, L., 2009. Factors Influencing Waste Separation and Utilization Among Households in the Lake Victoria Crescent, Uganda. *Waste Management*, 29, 3047-3051.

Gawaikar, V. and Deshpande, V.P., 2006. Source Specific Quantification and Characterization of Municipal Solid Waste – A review. *IE (I) Journal-EN*, 86, 33-38.

Gidman, P., Blore, I., Lorentzen, J. and Schuttenbelt, P., 1995. *Public-Private Partnerships in Urban Infrastructure Services. UMP Working Paper Series 4. Kenya, 1995.*

Halla, F. and Majani, B., 1999. Innovative Ways for Solid Waste Management in Dar-Es-Salaam. Toward Stakeholder Partnerships. *Habitat International*, 23 (3), 351-361.

Hedén, M., 2001. *Management of Household Waste in Samarinda Municipality, -Can we learn from Sweden?* East Kalimantan, Indonesia: Royal Institute of Technology.

Henry, K.R., Yongsheng, Z. and Jun, D., 2006. Municipal Solid Waste Management Challenges in Developing Countries – Kenyan Case Study. *Waste Management*, 26, 92-100.

Hockett, D., Lober, J.D. and Pilgrim, K., 1995. Determinants of Per Capita Municipal Solid Waste Generation in the Southeastern United States. *Journal of Environmental Management*, 45, 205-217.

Ilala Municipal Council, 2001. *Solid Waste Management Refuse Collection Fees Bylaws 2001.* Dar es Salaam, Tanzania.

Ishengoma, A., 2000. *Work from Solid Waste: Women Paid as Managers.* Solid Waste Management in Promoting Environmentally Sustainable Development Urban Development in Dar es Salaam City, The International Conference on Environmental and Social Perspectives for Sustainable Development in Africa. Arusha, Tanzania: Unpublished Paper, pp. 14.

Jenkins, R.R., 1993. *The Economics of Solid Waste Reduction. The Impact of Users Fees.* Brookfield, VT, USA: Edward Elgar Publishing Limited.

JICA, 1997. *Final Report – Executive Summary.* Dar es Salaam City Commission.

Joseph, K., 2006. Stakeholder Participation for Sustainable Waste Management. *Habitat International*, 30, 863-871.

Kachenje, E.Y., 2005. *Home-based Enterprises in Informal Settlements of Dar es Salaam, Tanzania.* Stockholm, Sweden: Kungliga Tekniska Höskolan, Royal Institute of Technology, 77 pp.

Kaseva, M.E. and Gupta, S.K., 1996. Recycling – An Environmentally Friendly and Income Generating Activity Towards Sustainable Solid Waste Management. Case study: Dar es Salaam City, Tanzania. *Resources, Conservation and Recycling*, 17, 299-309.

Kaseva, M.E. and Mbuligwe, S.E., 2005. Appraisal of Solid Waste Collection Following Private Sector Involvement in Dar es Salaam City, Tanzania. *Habitat International*, 29 (2), 353-366.

Kaseva, M.E., Mbuligwe, S.E. and Kasenga, G., 2002. Recycling Inorganic Domestic Solid Wastes. Results from a Pilot Study in Dar es Salaam City, Tanzania. *Resource, Conservation and Recycling*, 35, 243-257.

Kassenga, G.R., 1999. Potential and Constraints of Composting as a Market Solid Wastes Disposal Option for Dar es Salaam City in Tanzania. *Journal of Solid Waste Technology and Management*, 26 (2), 87-94.

Kassim, S.M., 2006. *Sustainability of Private Sector in Solid Waste Collection – A case of Dar es Salaam Tanzania.* Loughborough, UK: Loughborough, University, 347 pp.

Kiely, G., 1997. *Environmental Engineering.* London, UK: McGraw Hill.

Kinondoni Municipal Commission, 2000. *Waste Management and Refuse Collection Fees Bylaws 2000.* Dar es Salaam, Tanzania.

Kinondoni Municipal Commission. (2001). *Waste Management and Collection of Refuse Fee. By-laws GN. 354 of 2001.* Kinondoni Municipal Council, Dar es Salaam, Tanzania.

Kironde, J.M., 1999. Dar es Salaam Tanzania. In: A.G. Onibokun (ed.), *Managing the Monster: Urban Waste and Governance.* Ottawa, Canada: IDRC, pp. 101-172.

Kombe, W.J., 1995. *Formal and Informal Land Management in Tanzania: Case of Dar es Salaam City.* Dortmund, Germany: SPRING.

Kombe, W.J., 2001. Institutionalising the Concept of Environmental Planning and Management: Successes and Challenges in Dar es Salaam. *Development and Cities,* 11 (2 & 3), 190-207.

Kreuger, R.A., 1988. *Focus groups: A practical guide for applied research.* London, UK: Sage Publications.

Kreuger, R.A., 1994. *Focus Group. A Practical Guide for Applied Research.* Thousand Oaks, CA, USA: Sage.

Krishnaswami, O.R. and Ranganatham, M., 2007. *Methodology of Research in Social Sciences.* Delhi, India: Himalaya Publishing.

Kyessi, A., 2002. Community Participation in Urban Infrastructure Provision. Servicing Informal Settlements in Dar es Salaam. Spring Center, University of Dortmund.

Kyessi, A. and Mwakalinga, V., 2009. *GIS Application in Coordinating Solid Waste Collection: The Case of Sinza Neighbourhood in Kinondoni Municipality, Dar es Salaam City, Tanzania.* Dar es Salaam, Tanzania, pp. 19.

Kyessi, A.G., 2005. Community-Based Urban Water Management in Fringe Neighbourhoods: The case of Dar es Salaam, Tanzania. *Habitat International,* 29, 1-25.

Lardinois, I. and Van de Klundert, 1993. *Organic waste. Options for Small Scale Resource Recovery. Urban Solid Waste Series 1: Technology Transfer for Development (TOOL).* Amsterdam & Waste Consultants, Gouda, the Netherlands.

Letema, S., Van Vliet, B.J.M. and van Lier, J.B., 2010. Reconsidering Urban Sewer and Treatment Facilities in East Africa as Interplay of Flows, Network and Spaces. In: B. Van Vliet, G. Spaargaren and P. Oosterveer (eds.), *Social Perspectives on the Sanitation Challenge.* Dordrecht, the Netherlands: Springer, pp. 145-162.

Lewis, M., 2000. *Focus Group Interviews in Qualitative Research. A Review of the Literature.* Action Research Electronic Reader. Available: http://www.scu.edu.au/schools/gcm/ar/arr/arow/rlewis.html.

Linares, C.A., 2003. *Institutions and the Urban Environment in Developing Countries. Challenges, Trends, and Transitions.* Hixon Center for Urban Ecology: Yale School of Forestry & Environmental Studies.

Magrinho, A., Didelet, F. and Semiao, V., 2006. Country Report. Municipal Solid Waste Disposal in Portugal. *Waste Management,* 26, 1477-1489.

Majani, B., 2000. Institutionalizing Environmental Planning and Management. The Institutional Economics of Solid Waste Management in Tanzania. Spring Center. University of Dortmund.

Majani, B. and Halla, F., 1999. Innovative ways for solid waste management in Dar es Salaam: towards stakeholder partnerships'. *Habitat International,* 23 (3), 351-361.

Manga, V.E., Forton, O.T. and Read, A.R., 2008. Waste Management in Cameroon: A New Policy Perspective? *Resources, Conservation and Recycling.,* 52, 592-600.

Mbuligwe, S.E., 2002. Institutional Solid Waste Management Practices in Developing Countries: A Case Study of Three Academic Institutions in Tanzania. *Resource, Conservation and Recycling,* 35, 131-146.

Mbuligwe, S.E. and Kassenga, G.R., 2004. Feasibility and Strategies for Anaerobic Digestion of Solid Waste for Energy Production in Dar es Salaam City, Tanzania. *Resources, Conservation and Recycling,* 42, 183-203.

Medina, M., 2000. Scavenger Cooperatives in Asia and Latin America. *Resource, Conservation and Recycling,* 31, 51-69.

Ministry of Health, 2004. *National Environmental Health and Sanitation Policy Guidelines.* Dar es Salaam, pp. 67.

Mkwela, H. and Banyani, M., 2008. *Urban Farming: A Sustainable Solution to Reduce Solid Waste problems in Dar es Salaam, Tanzania.* Dar es Salaam, Tanzania: Ardhi University.

Morgan, D.L., 1993. *Successful Focus Groups: Advancing the State of the Art.* Newbury Park, CA, USA: Sage Publications.

Mosler, J.H., Drescherb, S., Zurbrügg, C., Rodri'guez, C.T. and Miranda, G.O., 2006. Formulating Waste Management Strategies Based on Waste Management Practices of Households in Santiago de Cuba, Cuba. *Habitat International,* 30, 849-862.

Mruma, O.R., 2005. *Implementation of the National Environmental Policy: A Case of Local Government Authorities in Dar es Salaam Tanzania.* Spring: University of Bergen,, 139 pp.

Mtshali, S.M., 2002. *Household Livelihood Security in Rural KwaZulu-Natal, South Africa*, Wageningen, the Netherlands: Wageningen University, 289 pp.

Mugagga, F., 2006. *The Public-Private Sector Approach to Municipal Solid Waste Management. How does it Work in Makindye Division, Kampala District, Uganda?* Trondheim, Norway: Norwegian University of Science and Technology (NTNU), 136 pp.

Mwai, M., Siebel, A.M., Rotter, S. and Lens, P., 2008. *Intergrating MDGs in the Formulation of Strategies for Solid Waste Management. A Life Cycle Approach.* WaterMill Working Paper Series no. 15.

NBS, 2002. *Tanzanian Household Budget Survey 2000/2001. National Bureau of Statistics (NBS), Dar es Salaam.* Dar es Salaam: Tanzania, National Bureau of Statistics, pp. 220. Available at: http://www4.worldbank.org/afr/poverty/databank/docnav/default.cfm.

NEP, 1997. *The National Environmental Policy.* Vice President Office, Dar es Salaam, Tanzania.

Niehof, A., 1999. *Household, Family and Nutrition Research. Writing a Proposal.* Wageningen H & C Publication Series 1, Wageningen University, the Netherlands.

Niehof, A., 2004a. The Significance of Diversification for Rural Livelihood Systems. *Food Policy*, 29, 321-328.

Niehof, A., 2004b. Why should households care? *Medische Antropologie*, 16 (2), 282-289.

Nkya, E., 2004. Public-Private Partnership and Institutional Arrangements: Constrained Improvement of Solid Waste Management in Dar Es Salaam. *Uongozi Journal of Management Development*, 16 (1), 1-21.

Nombo, I.C., 2007. *When AIDS Meets Poverty: Implications for Social Capital in a Village in Tanzania*, Wageningen: Wageningen University, 280 pp.

Nshimirimana, J., 2004. *Attitudes and Behaviour of Low-income Households Towards the Management of Domestic Solid waste in Tafelsig, Mitchell's Plain.* University of Western Cape, South Africa, 120 pp.

Ojeda-Benitez, S., Lozano-Olvera, G., Adalberto Morelos, R. and Armijo de Vega, C., 2008. Mathematical Modeling to Predict Residential Solid Waste Generation. *Waste Management*, 28, 7-13.

Okot-Okumu, J., 2006. *Solid Waste Management in Uganda: Issues Challenges and Opportunities.* Paper Presented at PROVIDE Programme Workshop, Wageningen, The Netherlands.

Oosterveer, P. and Spaargaren, G., 2010. Meeting Social Challenges in Developing Sustainable Environmental Infrastructures in East African Cities. In: B.J.M. van Vliet, G. Spaargaren and P. Oosterveer (eds.), *Social Perspectives on the Sanitation Challenge.* Dordrecht, the Netherlands: Springer, pp. 11-30.

Oteng-Ababio, M., 2009. Private Sector Involvement in Solid Waste Management in Greater Accra Ghana Metropolitan Area in Ghana. *Waste Management and Research*, 28 (4), 322-329.

Oyelola, O.T. and Babatunde, A.I., 2008. Characterization of Domestic and Market Solid Wastes at Source in Lagos Metropolis: Lagos, Nigeria. *African Journal of Environmental Science and Technology*, 12 (3), 43-437.

Parizeau, K., Maclaren, V. and Chanthy, L., 2006. Waste Characterization as an Element of Waste Management Planning: Lessons Learned from a Study in Siem Reap, Cambodia. *Resource, Conservation and Recycling*, 49, 110-128.

Parrot, L., Sotamenou, J. and Kamgnia Dia, B., 2009. Municipal Solid Waste Management in Africa: Strategies and Livelihoods in Yaoundé, Cameroon. *Waste Management*, 29, 986-995.

Pflanz, M., 2006. Tanzania to Ban all Plastic Bags. *The Daily Telegraph*.

Post, J. and Baud, I., 2004. Government, Market and Community in Urban Solid Waste; Problems and Potentials in the Transition to Sustainable Development. In: I. Baud, J. Post and C. Furedy (eds.), *Solid Waste Management and Recycling.* Dordrecht, the Netherlands: Kluwer Academic Publishers, pp. 259-281.

Qu, X., Li, Z., Xie, X., Sui, Y., Yang, L. and Chen, Y., 2009. Survey of Composition and Generation Rate of Household Wastes in Beijing, China. *Waste Management*, 29, 2618-2624.

Reem, M., 2002. *Solid Waste Policy Making in a System in Transition. The Case Study of Biological Treatment in the West Bank Institute of Community and Public Health.* Birzeit University, Palestinian territories.

Sawio, C.J., 2008. Perception and Conceptualisation of Urban Environmental Change: Dar es Salaam City. *Geographical Journal*, 174 (2), 149-175.

Scheinberg, A., Muller, M. and Tasheva, E., 1998. *Gender and Waste: Integrating Gender into Community Waste Management.* Gouda, the Netherlands: WASTE.

Schübeler, P., Wehrle, K. and Christen, J., 1996. *Conceptual Framework for Municipal Solid Waste Management in Low-Income Countries. Collaborative Programme On Municipal Solid Waste Management in Low-Income Countries.* SKAT/UNDP/UNCHS (Habitat)/World Bank/SDC.

Simon, A.M., 2008. *Analysis of Activities of Community Based Organizations Involved in Solid Waste Management, Investigating Modernized Mixtures Approach. The Case of Kinondoni Municipality, Dar es Salaam, Tanzania.* Wageningen, the Netherlands: Wageningen University, 93 pp.

Šliužas, R.V., 2004. *Managing Informal Settlements. A Study Using Geo-Information in Dar es Salaam, Tanzania.* Utrecht, the Netherlands: Utrecht University and ITC, 327 pp.

Snel, M. and Ali, M., 1999. *Stakeholder Analysis in Local Solid Waste Management Schemes.* Task No:69. Water and Environmental Health at London and Loughborough (WELL).

Spaargaren, G., Oosterveer, P., van Buuren, J. and Mol, A.P.J., 2005. *Mixed Modernities: Towards Viable Urban Envirtonmental Infrastructure Development in East Africa.* Wageningen Environmental Policy Group, Wageningen. Wageningen University.

Stewart, D.W. and Shamdasani, P.N., 1990. *Focus Groups: Theory and Practice.* Newbury Park, USA: Sage Publications.

Stewart, D.W., Shamdasani, P.N. and Rook, D.N., 2007. *Focus Groups: Theory and Practice. Applied Social Research Methods*, 20. Thousand Oaks, CA, USA: Sage Publications.

Tadesse, T., Ruijs, A. and Hagos, F., 2008. Household Waste Disposal in Mekelle City, Northern Ethiopia. *Waste Management*, 28 (10), 2003-2012.

Tchobanoglous, G., Theisen, H. and Vigil, S.A., 1993. *Integrated Solid Waste Management. Engineering Principles and Management Issues.* New York: McGraw-Hill, International Editions.

Temeke Municipal Council, 2000. *Waste Management and Refuse Collection Fees Bylaws 2000.* Dar es Salaam, Tanzania.

UN-HABITAT, 2004a. The Demonstration City. *The SCP Documentation Series. The Sustainable Dar es Salaam Project 1992-2003. From a City Demonstration Project to a National Programme for Environmentally Sustainable Urban Development.* Nairobi, Kenya: UN-HABITAT, pp. 23-51.

UN-HABITAT, 2004b. The SDP Dar es Salaam Project, *The SCP Documentation Series, Project 1992-2003. From Urban Environment Priority Issues to Up-Scaling Strategies city-wide.* Nairobi, Kenya: UN-HABITAT, pp. 4-20.

UN-HABITAT, 2007. *Situation Analyisi of Informal Settlements in Kampala.* Nairobi, Kenya.

UN-HABITAT, 2009. *Tanzania. Dar es Salaam City Profile.* Nairobi, Kenya.

UN, 2004. *United Nations Demographic Yearbook review.* New York, NY, USA: UN.

UNCED, 1992. *Report of the UnitedNations Conference on Environment and Development. Rio de Janeiro, 3-14 June 1992.* New York, NY, USA: United Nations.

UNCHS, 1994. *Sustainable Cities Programme. Concepts and Applications of a United Nations Programme.* Nairobi, Kenya: UNCHS.

UNEP, 2009. *Developing Integrated Solid waste Management Plan Training Manual, Volume 2. Assesment of Current Waste Management System and Gaps Therein.* Osaka/Shiga, Japan.

United Nations, 2006. *World Urbanisation Prospects: The 2005 Revision. New York: Department of Economic and Social Affairs, Population Division. Available at:* http://esa.un.org/unup/.

United Repbulic of Tanzania, 1997. *National Environmental Policy*: Vice President's Office, pp. 6-8.

URT, 2003. *Waste Management Guidelines.* Ministry of Health, Dar es Salaam, Tanzania.

URT, 2009. *National Public Private Partnership (PPP) Policy.* Ministry of Health, Dar es Salaam, Tanzania.

USEPA, 2002. *What is Integrated Solid Waste Management? United States Environmental Protection Agency.* Washington, DC, USA: USEPA.

Van de Klundert, A. and Anschütz, J., 2000. *The Sustainability of Alliances Between Stakeholders in Waste Management. Working Paper for UWEP/CWG – Draft*, UMP Workshop on Municipal Solid Waste Management, Ittingen, Switzerland.

Van de Klundert, A. and Anschütz, J., 2001. Integrated Sustainable Waste Management – the Concept. In: A. Scheinberg (ed.), *Tools for Decision-Makers. Experiences from the Urban Waste Expertise Programme (1995-2001).* Gouda, the Netherlands: WASTE.

Van de Klundert, A. and Lardinois, I., 1995. *Community and Private (Formal and Informal) Sector Involvement in Municipal Solid Waste Management in Developing Countries.* Background Paper for Discussion at the 'Ittingen Workshop' Jointly Organised by the Swiss Development Cooperation and the Urban Management Programme.

Van de Walle, E., 2006. *African Households: Censuses and Surveys (General Demography of Africa).* Portland, OR, USA: M.E. Sharpe, 247 pp.

Van Horen, B., 2004. Fragmented Coherence: Solid Waste Management in Colombo. *International Journal of Urban and Regional Research*, 757-773.

Van Koppen, C.S.A., 2004. Eco-industrial Design of Biomass Cycles. Flows, Technologies, and Actors. In: Lens, P., Hamelers, B., Hoitink, H., Bidlingmaier (eds.) *Resource Recovery and Reuse in Organic Solid Waste Management (Integrated Environmental Technology).* London, UK: IWA Publishing, pp. 24-43.

Venkatachalam, P., 2009. *An Overview of Municipal Finance System in Dar es Salaam, Tanzania. Crisis state occasional papers.* London, UK: LSE Destin. Available at: http://129.132.57.230/serviceengine/Files/ISN/109983/ipublicationdocument_singledocument/f57d25c8-bb95-46e0-838e-4b4541db405d/en/OP10Venkatachalam.pdf.

Weber, R.P., 1990. *Basic Content Analysis. Quantitative Applications in the Social Sciences.* Newbury Park, CA, USA: Sage.

Wilson, C.D., Velis, C. and Cheeseman, C., 2006. Role of Informal Sector Recycling in Waste Management in Developing Countries. *Habitat International*, 30 (4), 797-808.

World Bank, 2002a. *Data by Country. Available at:* http://www.worldbank.org/data/countrydata/country data.

World Bank, 2002b. *Upgrading Low-income Urban Settlements.* Country Assessment Report, Tanzania.

Yhdego, M., 1993. Composting Organic Solid Waste in Dar es Salaam City. *Resources, Conservation and Recycling*, 12, 185-194.

Yin, R.K., 2003. *Case Study Research and Methods. Applied Social Research Methods Series*, 5. Thousand Oaks, CA, USA: Sage Publications.

Yusof, M.B., Othman, F., Hashim, N. and Ali, C.N., 2002. The Role of Socio-Economic and Cultural Factors in Municipal Solid wastw generation: A Case Study in Taman Perling, Johor Bahru. *Jurnal Teknologi*, 37, 55-64.

Zerbock, O., 2003. *Urban Solid Waste Management. Waste Reduction in Developing Nations* Michigan Technological University, USA, 23 pp.

Žičkienė, S., Tričys, V. and Kovierienė, A., 2005. Municipal Solid Waste Management. Data Analysis and Management Options. *Environmental Research, Engineering and Management*, 3 (33), 47-54.

Appendix 1. Households survey questionnaire with regard to solid waste management in Dar es Salaam City – Tanzania

A. Purpose of the questionnaire

Can we have a moment of your time to assist complete a survey on solid waste management at your household. We would like to ask you some questions that would assist us in determining how to improve solid waste services within households. This is part of a PhD research. Your household is among the selected households for interviews, so please your contribution is considered very valuable to this survey. (All answers will be treated confidentially).

B. Identification

Enumerator name ....
Interview start .... Interview end ....
Date...
Name of respondent/household head ....

Part 1: Demographic information

1. Location and size of the household.

	Municipality	Ward	Sub-ward
Plot/house no.			
Number of people in the household – adults and children			

2. Characteristics and status of respondents

Head of household	Sex	Marital status	Age
a. Yes	a. Male	a. Married	
b. No	b. Female	b. Single	
		c. Widow	

3. Socioeconomic variables

Education	Source of income	Monthly income	Occupancy status
a. Primary	a. Formal	a. 0-50,000	a. Owner
b. Secondary	b. Informal	b. 50,001-100,000	b. Tenant
c. Tertiary	c. Retired	c. 100,000-150,000	c. Cohabiting
d. None	d. Other	d. 150,001-200,000	

4. Does your household has a business activity?
   a. Yes
   b. No
   If answer is (a). What kind of business? ....

Part II: Respondents knowledge on solid waste management

5. Do you know what solid waste is?
    a. Yes
    b. No

6. Have you heard or received any training on solid waste management?
    a. Yes
    b. No

7. If no 6 is (a), how did you get the training?
    a. Through Municipal health officer
    b. Through waste (CBO)/private company
    c. Through media
    d. Others

8. On what aspect of SWM were you trained or heard?
    a. Generation
    b. Waste handling
    c. Separation of waste
    d. Others

9. Do you know what a solid waste transfer station is?
    a. Yes (go to 23)
    b. No

Part III. Solid waste management: practices and household's daily routine

10. Is there a solid waste container in your household
    a. Yes (go to 11)
    b. No

11. Which type of container?
    a. Metallic bin
    b. Plastic bucket
    c. Plastic bags
    d. Other

12. Do you separate waste that is generated in your household?
    a. Yes (go to 13)
    b. No

13. Which items do you separate?

....

14. What do you do with items that you separate/recycle?
    a. Sell to waste collectors
    b. Own reuse
    c. Give it away to others who will use it again
    d. Others

15. Explain the reason for not separating waste

....

16. When does the general cleaning inside and outside to your house takes place?
    a. 5.00-6.00 a.m.
    b. 7.00-10.00 a.m.
    c. Any time
    d. Others

17. Where do you keep waste that you sweep from your house?
    a. Discard them with other solid waste
    b. Throw it in yard
    c. Throw in road sides
    d. Others

18. Who is responsible for handling the waste from the house?
    a. Mother
    b. Father
    c. Housemaid/children
    d. Others

Part IV: Service provision: collection and disposal

19. Who has the primary responsibility for collecting your household's solid waste?
    a. Municipality
    b. Waste contractor
    c. Not collected
    d. Don't know

20. How does your household cooperate in the collection of solid waste?
    a. Household member bring waste to a transfer station/waste truck
    b. There is no cooperation
    c. Other, please specify

21. If your waste remain uncollected, how do you discard it?
    a. Burn it
    b. Bury it in the backyard
    c. Throw it on the street
    d. Other, please specify

22. How many times per week is your solid waste collected from your house?
    a. Once a week
    b. Twice a week
    c. Now and then
    d. Other

Part V: Respondent's general views on solid waste management

23. Do you need a transfer station in your neighbourhood?
    a. It is important to have in the neighbourhood
    b. They produce unpleasant smell
    c. There is no need of it
    d. Don't know

24. What do you consider the most severe problem relating to management of solid waste?
    a. Public health risk
    b. Bad odour
    c. Nothing is wrong

25. What time would you like your waste to be collected from your house?
    a. Between 6.00 a.m. and 9.00 a.m.
    b. Between 9.00 a.m. and 12.00 p.m.
    c. Between 13.00 p.m and 18.00 p.m
    d. Any time

26. Which role would you like to play most in SWM?
    a. By paying waste fee
    b. Bring waste to communal facility (transfer station)
    c. Separate waste
    d. Not willing to do anything

27. What problems/constraints do you face in solid waste management within your household?

Thank you very much for your valuable time.

Appendix 2. Checklist format for observation units at household level in Dar es Salaam

Researcher and research assistants should record their observations in the table below

	Items to be observed		Hh1	Hh2	Hh3	Hh4	Hh5	Hh6	Comments or remarks
		Date							
		Time							
		Sub-ward							
1.	**Presence of different types of waste:**								
	a. Kitchen waste (food remains, vegetable and fruit peels (etc.)								
	b. Sweepings								
	c. Papers: magazines, newspapers, cardboards								
	d. Glass								
	e. Metals								
	f. Plastics								
	g. Residues								
2.	**Handling of solid waste: storage, collection and transfer**								
	Is waste stored at the household level? a. Yes b. No								
	If YES in which type of container is waste stored?								
	1. In bags?								
	2. In buckets?								
	3. In bins?								
	4. Any other (specify)								
	If NO, how is generated waste managed?								
	1. Burnt								
	2. Buried								
	3. Thrown in nearby drains/ street								
	4. Taken to TS								
	5. Collected by waste contractors								
	6. Others (specify)								

3.	**Is generated waste stored in one container or separate?**									
	a. One container									
	b. Separate									
	Where is the waste container placed?									
	c. Kitchen									
	d. Yard of the house									
	e. Others (specify)									
4.	**Is there a container for separated waste?**									
	a. Yes									
	b. No									
	If YES, what happens to separated waste?									
	1. Re-used at household level									
	2. Sold									
	3. Mixed with other waste									
	4. Recycled									
	5. Composted									
5.	**If waste is composted how is it done?**									
	a. In a pit									
	b. Open air									
	c. In a container like a box									
	d. Others									
6.	**Who is responsible for managing waste?**									
	a. Father									
	b. Mother									
	c. Housemaid									
	d. Boy child									
	e. Girl child									

Appendix 3. Observation checklist at solid waste: transfer stations

Observed item	TS 1	TS 2	TS 3	TS 4	TS 5	TS 6
1. Type of facility						
a. Transfer station:						
- Permanent						
- Movable						
b. Collection points:						
- Open space						
- Side of the street						
- Standby trailer						
c. Bins						
2. Location: distance to households						
a. Within households premises						
b. <50 m						
c. 50-100 m						
d. 100-500 m						
e. >500 m						
3. Who is bringing waste to the transfer station						
a. Women						
b. Men						
c. Children						
4. Status of waste						
a. Separated						
b. Mixed						

Appendix 4. Interview guide for Municipal/City Officials in solid waste management – East African capital cities

Introduction

1. Department..
2. Position ....
3. Address ....

          Date ....
          Enumerator ....
          Time start ....
          Time finish ....

Solid waste management activities

- What are the sources of solid waste in your municipality?
- How much solid waste is generated daily in your municipality? What is the generation rate?
- What does the municipality do to promote or carry out in household waste management?
- Who is responsible for?
  1. Primary collection
  2. Secondary collection

Rules and regulations

- Do you have any municipal by-laws on SWM?
- What does the municipal by-laws state on the SWM in general concerning the following items:
  ○ Containers
  ○ Material of the container
  ○ Type of the container
  ○ Waste producer responsibility
  ○ Separation of waste at source
  ○ Liability for failure to pay for waste collection fees

Stakeholders in SWM

- What is the role of municipal officials in solid waste management in the municipality?
- Are there any other institutions or agencies which provide solid waste management services in your municipality? (find about private waste contractors and their service to households in informal settlements)
- Do they have any terms of contract with the municipality?
- What are their roles?
- What is the role of (1) households, (2) ward leaders, (3) counsellors, (4) sub-ward leaders in SWM? (5) Province, (6) Districts?
- What are municipal/city views regarding the roles in households waste management?

Institutional support

What kind of support does the municipality give to household waste management?

Equipment

- Which equipment/tools do you have for managing solid waste?
- What type of technological options are promoted by the municipality in household solid waste management?

Financial

How does the municipality assist household waste management financially?

Human resources

- Does the SWM department staff have job description detailing their daily tasks to household waste management?
- Does the municipality conduct any training and public awareness programs on SWM? Yes/No

Future plans

- What are the municipal long term plans on household SWM?
- What problems/constraints do you face in solid waste management at household level?
- What are the future plans/strategies the municipality has with regard to improving solid waste management at household level?

Thank you!

Appendix 5. Questions for interviews – service providers (CBOS/private companies)

1. Name ....

2. Position ....

3. Address ....

Interview start ....

Interview end .... Date ....

Explanation of our interest in SWM (you may say our interest is in households in low-income areas in informal settlements)

Background

• Who started the CBO or the company, and when?
• What were their motives in starting it?
• Why this particular type of organization (e.g. CBO? Or private company)

Structure of organization:

• What are its main activities in relation to SWM?
• Who is involved – how many members?
• How many employees do you have?
• What kind of employees? Permanent or temporary?

Service provisioning

• Which service do you provide in relation to solid waste management?
• Where do you provide the service? (e.g. what sub-ward, ward, province, etc.)
• What is the status of area that you provide the service? (try to probe issues relating to low-income households)
• How many households do you give the service?
• How do you collect waste from households? (method of collection)
• What type of tools and equipment do you use to provide SWM services? (find out the number of each equipment)
• What is the status of waste you collect from households? (sorted or mixed)
• Where do you take the waste that you collect?
• Is there transfer station in the area that you provide service?
• Do you take waste to transfer station? (who transfer it)
• What is your collection schedule? (time, frequency of collection)
• Do households pay for the collection services? (mode of payment – how do you collect waste fee)

Municipal support

- Do you get any municipal support in servicing households?
- What kind of support? (financial, technological, human)

Particular issues for households

- Is there any problem you encounter in providing services to households?
- How is households involvement awareness?
- From your own opinion what are your suggestions with regard to improving the provision of this service to households?

Thank you!

Appendix 6. Household solid waste management in East African capital cities: facilitator's guide for focus group discussion

Total participant's time required: 2 hours
Total focus group time: 1.95 hours
Total break time: 0.05 minutes

I. Introduction (5 minutes):

a. Moderator introduces her/himself and explains the purpose of the meeting

Good morning/good afternoon. My name is _____. I am working at _____.
We are working on a project for the solid waste management on matters concerning solid waste in our households. First of all let me thank you for the time you find to come here today. You have been requested to participate because we believe that you can give us ideas and views on the issues related to solid waste in our households.

b. Explain focus group process

A focus group is a research method for collecting data where questions are posed to the whole group and everyone is asked to respond and talk to each other. We are interested in your own opinions, in other words, what you think and feel about each topic of discussion. I think no one of you have participated before in such a group discussion we are going to initiate just now. Today the subject matter of our meeting is to discuss the situation with household solid waste management in our households.

c. Purpose of the discussion

The purpose of our discussion is to find out what do you personally think about household waste management. I am acting here not as an expert or specialist for the matters in question intending to teach you something. On the contrary, you yourselves will be the experts today, and, therefore, I would like to hear what are your personal ideas. The results of study will be further used to change the situation with solid waste management in our households.

d. Discussion procedure

Let me explain some of the basic rules for the conduct of our discussion:
- First, our discussion will take approximately two hours. I want to assure you that no one, else, would ever hear anything you will say here. Prior to preparation of a report, your opinions will be summarized, and your names will be never mentioned anywhere.
- Further, we would like to hear your opinions regarding every issue which will be discussed from the viewpoint of your own experience. During discussions no answers would be treated to be incorrect, therefore, please, feel absolutely free to say everything you think.

- Third rule. It is very important for us to hear each of you. You are absolutely not required to answer each question, but, please, try to express your thoughts any time you have something to say. Besides, let other participants speak out as well. Also, you are kindly requested to avoid discussing any other matters unrelated thereto so that all of you could hear well enough of what we are talking about.
- And finally, the last item. I will not express my own opinion. My role is to run the discussion so that every one of you may get a possibility to speak out, and make myself certain of the fact that all matters in question have been discussed. So, any questions from you?

e. Introduction of participants

Let us start with introducing each other. Please, tell your names, where you come from. I'll start with myself. (introduce yourself)

II. Main problems (15 minutes)

All of you probably know and feel that in our households there are a lot of problems relating to environment. With this in view, I would like to speak a little about those problems which you believe are most vital for today.
- Please, indicate environmental problems which, to your opinion, is the most important for us in our area.
- It is critical for me to not only hear from you that this or that problem is the most important, but I would like you to substantiate somehow your opinion. (try to Focus on issues related to solid waste management)
- What can you say about solid waste management in our households?
- What do you consider the most urgent problem related to solid waste management in your household? (try to clarify in case the participants do not mention the problem as the most important)

III. Household solid wastes management practices (80 minutes)

Now I would like to discuss the situation of wastes management at your houses. Please, try to speak out in more details about everything that might be related to this issue.

a. Generation and storage of solid waste at households

- Please mention different types of waste that is generate at your household.
- Please, describe how you usually store the waste at your household. Do you have any special containers for waste storage? How many containers do you have for waste storage? If more than one, why so?
- Where do you place your waste container? Indoor? Outdoor?
- Do you face any problem with waste storage at your household? What are these problems? How do you manage to settle such problems?

b. Waste separation

Now let us speak a little of whether you separate solid waste at your household, and how do you do it? I would like first to ask whether you know what is meant by waste solid waste separation.
- Do you separate solid wastes to some portions? What exactly are those portions? Why to such portions? What do you do with the separated portions of solid wastes?
- Have you ever left anything for solid wastes with you so as to re-use those? Namely, what? Why?
- What do you do with your kitchen waste? Composting? Mix with other waste? (to investigate matters in respect to glass, plastics and papers, foodstuff wastes)

c. Collection and disposal of household waste

Now I would like to discuss how the collection and disposal of solid waste is done in your household. We should discuss where your solid waste is taken away to, who removes it and when? I would also like you to remember as many details as possible from your everyday life. So, let us start with the question how is waste collected from your households? Please, describe in more details. (investigate the issues related to open dumping, burning, burying, etc.)
- Do you face any problems with solid waste collection at your households? What are these problems? How do you manage to settle such problems?
- Who is usually in charge to bring out the solid waste at your household? Is there a member of your household who is assigned with this duty?
- How often is solid wastes collected from your household? What are the vehicles arriving to pick up the solid waste – specialized or usual loaded with solid waste by manpower? Wheelbarrows? Pushcarts?
- Who, is responsible for the waste disposal? Solid waste contractor? Informal picker? Myself bring to transfer station? Municipality?
- Whether the solid waste collected nearby your household is burnt? How often it happens? Who sets, as to your observations, the waste on fire – those who are responsible for its waste collection, or just playing children?
- Do you know where the solid waste collected from your household transported to? Please, indicate the specific place, if you know. How far is this place from your place?

d. Payment for solid waste collection

Now I would like to discuss some issues related to solid waste collection fees.
- How much do you pay per month for solid waste collection? Whether you usually pay every month or sometimes the payment is delayed by you?
- What is the procedure of payment for solid waste collection? Where do you usually pay for solid waste collection? Is such payment system convenient? Why 'not', and why 'yes'?

e. Alternative methods of solid waste collection

As to your opinion, whether the whole process of solid waste collection and waste collection could be arranged in some other way? How could it be done? (to investigate about possibility of households to transfer waste to transfer station, etc.)

IV. How to contribute for improvements (20 minutes)

So, we have discussed together the entire process of solid waste management from your households. We have only remained to discuss what kind of information might be needed for us, what could be likely improvements in the sector of solid waste collection, and your willingness to participate in such improvements. It is natural that you may be involved thereto only in one way – to pay more RCC and transportation of solid waste.

- How would you be willing to contribute for improvement? Bring your own waste to a transfer station or communal facility? Are you ready to pay more per month provided that solid waste collection would become better, i.e. more clean and regular, etc.? Why 'yes', or why 'not'? Separate your waste? Or you are not willing to do anything?
- Whom you would prefer to carry out waste separation/recycling/composting – solid waste contractors, municipality or households? Informal picker? Why?

V. Cooperation with waste contractors/municipal officials

We have discussed your willingness for improvement of solid waste management. Let's hear from your opinions how would you prefer to cooperate with waste contractors and the municipal officials. How do you cooperate with solid waste contractors and the municipal officials in solid waste management? (to investigate roles of: municipality, solid waste contractors, perceptions on household role, etc.)

IV. End of discussion

That is all with our discussion. Thank you very much for making the time available to participate. Please, receive your remuneration.

Appendix 7. Organization chart for the Department of Environment – Nairobi, Kenya

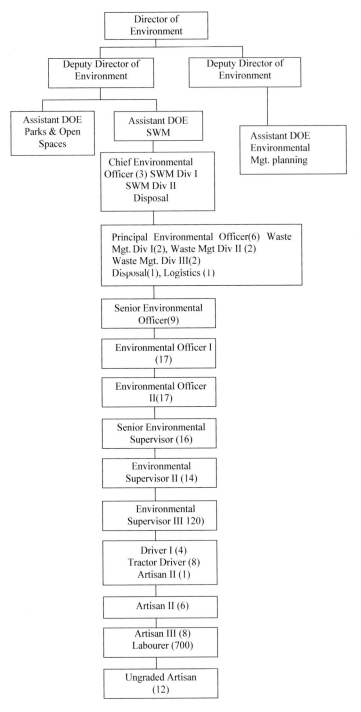

Appendix 8. Decentralized environmental management framework in Kampala, Uganda (KCC, 2009)

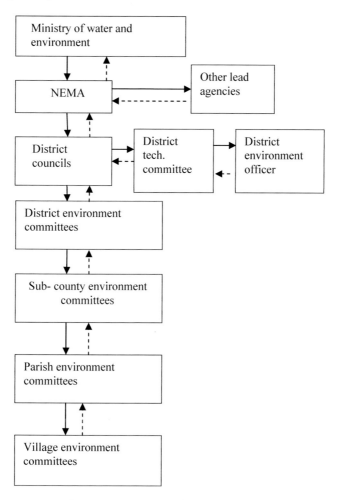

Note: Continuous arrow = coordination, supervision and capacity building; broken arrow = reporting and information flow.

Summary

This thesis analyses the different roles households fulfil in urban solid waste management (SWM) chains. The study targets in particular households living in the high density informal settlements of East African capital cities. So far, in-depth understanding of how householders deal with solid waste within their household and how they link with other actors in the chain is lacking while this is necessary to introduce improvements.

The central research questions in this study are:
1. What are the characteristics of household wastes?
2. What are the key domestic practices and routines for handling solid wastes in the primary phase, at the household level?
3. How do households receive solid waste management services and what are their basic perceptions and evaluations of the way of being served?

The study starts with developing a conceptual framework based on the Modernized Mixture Approach, to allow for the identification of appropriate conceptual tools to understand the roles of households in SWM chains. Three key roles are identified: households as waste generators, as waste handlers in the primary phase of the chain and as recipients of municipal solid waste services.

These different roles were analysed through a comprehensive empirical study covering low-income households in Kinondoni municipality in Dar es Salaam, Tanzania. Multiple data collection methods were employed and comprised semi-structured interviews with key informants, a household survey, direct observation, and focus group discussions. The household survey was carried out along with a waste characterization study among 360 respondents (households) sampled from 6 sub-wards of the Kinondoni Municipal Council. In addition, the relevant policy documents were reviewed to substantiate the information derived from interviews. To analyse the role of households as waste generators, the per capita generation and composition of household solid waste and the factors influencing this were empirically investigated. For studying the role of households as waste handlers, the field study entailed investigating their management practices, the solid waste flow from the households to the transfer station, the alternative methods that households use in managing waste, and the roles of different household members in waste management. With regard to the role of households as service recipients the study looked at the formal and informal stakeholders in households' SWM, the relationships existing between households and other stakeholders, the kinds of institutional support that exists for households and other stakeholders and the households' perceptions. Finally, these findings from Dar es Salaam were compared with findings from Nairobi (Kenya) and Kampala (Uganda) to put them in a wider perspective.

The results of the household's waste characteristics study showed that the overall average of the per capita waste generated across the selected sample was 0.44 kg/day. When differentiated according to socio-economic characteristics, the per capita waste generated varies from 0.36 kg/day in the low-income sub-wards to 0.52 kg/day in the middle income ones. On average kitchen/food waste stood out as the major (70%) household waste component. We observed that the fractions which constituted kitchen/food waste originated not only from the normal preparation of the households' own meals, but also from food related businesses conducted within these households.

In line with this, our survey indicated that 52% of the interviewed households carried out food-related businesses. The factors contributing to the quantity and composition of household solid waste were found to relate to daily cleaning routines, daily economic activities of households, recovery of recyclable materials within the households and to cooking and eating habits. As food wastes are mostly of organic nature the possibilities of recycling through composting deserve further analysis.

The analysis of the role of households as waste handlers in the SWM chain focussed on practices within the households and on the flow of generated waste from the household to the transfer station (or other modes of disposal). The observations showed that households practice waste sorting without the knowledge of and probably also without recognition from the municipal officials. Households sort out the materials which they consider useful for purposes such as storing oil, medicine etc. without further knowledge of the environmental principles and purposes of waste sorting. These practices are motivated rather by economic (scarcity) motives than by environmental considerations. However, also waste separation for technical or economic reasons should be considered in future policies on domestic SWM in informal urban settlements. With regard to waste containers our empirical findings revealed that households consider storage containers as facilities without any value, so as waste too. This means that households eventually dispose of these waste containers together with the wastes they stored. Households want a clean environment and they consider SWM a housekeeping duty, perceived as the domain of housemaids and children under the supervision of women. Women are expected to execute and supervise the household work including taking care of solid waste. This implies that SWM is given a low status at household level with male adults not taking any responsibility. These social and cultural factors are important constraining factors for realizing more sustainability in domestic SWM by households.

In analysing the waste-flows in between households and transfer stations, we observed that the skip (or container or transfer station), is important in the SWM chain. The transfer point not only receives wastes from households but also functions as the connection between the primary and secondary phase of the SWM chain. In the pre-transfer station phase, householders and their everyday life rationalities and practices of storing, (non)separating and delivering wastes are dominant. The secondary phase of the waste chain is dominated by formal municipal institutions and private corporations applying a system-based rationality. The transfer station constitutes an important focal point from households' perspectives, so SWM practices by households and (formal and informal) service providers are structured around it.

As recipients of solid waste services households have become embedded in SWM chains through both formal and informal relationships. In formal relationships, households are mostly passive recipients of services, placed at the down-stream of the chain, with the municipality being at the top, assigned formal authority to organize, monitor and supervise service delivery. In this structure, the municipality is recognized as the key regulator, whereas waste contractors are recognized as actual providers when registered and given a license by the municipality. The responsibility of the municipality to the households is to make sure that they receive SWM in order to avoid the outbreak of diseases and to protect the environment in general. Informal relationships exist as well through a collaboration between households and illegal waste pickers which is not officially recognized by municipal officials. Informal relationships can also exist between households and official waste contractors as a result of mutual agreements on issues which are not covered by the municipal by-laws. Both formal and informal relationships exist in parallel, complementary or in competition with each other.

In comparing our findings in Dar es Salaam with similar data from informal settlements in Nairobi and Kampala we focussed on service provisioning (the main actors in SWM), waste handling (the flow of waste to the transfer station) and the perception of households regarding the services they receive. Regarding service provisioning, we found that privatization of SWM services has allowed the involvement of other stakeholders (private waste contractors) in providing solid waste collection and disposal services that were traditionally considered the government responsibilities. With respect to waste handling (storage, collection and transfer), the three cities are similar and none of local authorities recognizes the importance of resource recovery and waste separation at household level. Regarding the perceptions of households on the SWM services provided we concluded that in all three cases households were concerned about the health risks and environmental problems resulting from inadequate solid waste collection and disposal services. Households consider the local municipal council responsible for providing adequate services. They raised a number of complaints on SWM and made suggestions for improving the present situation. As for their own roles in SWM, the householders express a willingness to contribute under the condition that also local authorities perform their roles and take their responsibilities while recognizing and considering the demands and concerns of householders.

Overall, this thesis concluded first that both technical and social factors play a role in household SWM. Lack of proper equipment, lack of human resources, inaccessibility of particular locations for trucks, lack of storage space, absence of proper (storage, separation, removal) technologies, the unacknowledged role of household within SWM are all very relevant factors in explaining the malfunctioning of SWM systems in informal settlements. Secondly, both formal and informal relationships are important. If formal waste management systems do not deliver services at an adequate level, informal practices and actors come into play. Once these informal activities are introduced, they tend to turn into established practices with stakeholders and specific interests attached to them. As our study has shown, informal relations and activities are indispensable elements of waste management systems in informal settlements and they need to be recognized and dealt with when trying to improve the functioning of these systems in the short run. Thirdly, households as key actors in SWM should be recognized for doing things their own way. In order to effectively improve SWM by designing new interventions, the specific nature and dynamic of household practices have to be taken into account. Fourth, solutions are not one of a kind but 'mixtures' that should be based on local circumstances. In agreement with the Modernised Mixtures Approach, this research confirmed that the situation of structural underperforming waste management systems in informal settlements in Dar es Salaam and other East African capital cities represents a complex, multidimensional problem that cannot be solved by one particular actor or strategy alone and that a mixture of actors and strategies is required.

About the author

Aisa Oberlin Solomon, PhD candidate in the Environmental Policy Group of Wageningen University, was born in Hai district, Kilimanjaro region in Northern Tanzania. She holds an Advanced Diploma in Public health Engineering from Ardhi Institute (1985) now Ardhi University (ARU), and a MSc degree in Urban Environmental Management from Wageningen University and Research Centre/Institute of Urban and Housing and Development Studies (2001). Aisa is a lecturer at Dar es Salaam Institute of Technology in the Civil Engineering Department. She teaches undergraduate courses related to water supply engineering, waste water engineering and solid waste management. In her career she has been involved in carrying out consultancy jobs related to sanitation and solid waste management. From 2002 to 2006 she coordinated a link on Solid Waste Management for Small and Medium Enterprises, part of the higher education links for the institutions in the UK and overseas, supported under Development Partnerships in Higher Education (DELPHE) project between Dar es Salaam Institute of Technology (DIT) and Leeds Metropolitan University (Leeds met). This link was funded by the UK Department for International Development (DFID), it aimed at contributing to reducing poverty, promoting science and technology and help achieving the UN Millennium Development Goals by 2015. From 2003 to 2009, with the International Labour Organization (ILO), Dar es Salaam Area Office for Eastern Africa she participated in conducting the trainings on Integrated Solid Waste Management (ISWM) with an entrepreneurship perspective to solid waste contractors providing solid waste management services in municipalities of Tanzania, Kenya and Uganda. The training aimed at improving solid waste management and other urban services whilst creating more and better jobs for the poor. She has held some positions in community services. She has been a Member of Olympio Primary school parents committee in Dar es Salaam, January 2003 to 2006; Member of DIT workers council, from August 2005 to 2006; Member of the executive committee of the DIT workers council, September 2005 to 2006; and board member of Zanaki girls secondary school in Dar es Salaam from 2008 to date. In November 2006 she obtained a Sandwich scholarship through the PROVIDE (Partnership research of viable infrastructure development) project to pursue her PhD with the Environmental Policy Group at Wageningen University and Research Centre. Her research areas of interest include solid waste management, sanitation and water supply.

She is married and mother of one son and two daughters.

Contact: aisa_oberlin@hotmail.com

Printed in the United States
by Baker & Taylor Publisher Services